A. M. Gibson

Report on the Coal Measures of Blount Mountain

With Map and Sections

A. M. Gibson

Report on the Coal Measures of Blount Mountain
With Map and Sections

ISBN/EAN: 9783337249724

Printed in Europe, USA, Canada, Australia, Japan

Cover: Foto ©berggeist007 / pixelio.de

More available books at **www.hansebooks.com**

GEOLOGICAL SURVEY

—OF—

ALABAMA,

EUGENE ALLEN SMITH, Ph. D., State Geologist.

REPORT

ON THE

COAL MEASURES

OF

BLOUNT MOUNTAIN,

WITH MAP,

AND SECTIONS

BY

A. M. GIBSON, ASSISTANT.

THE BROWN PRINTING CO., PRINTERS AND BINDERS.

TABLE OF CONTENTS.

Sir—In the following pages please find the results of the examination of the Blount Mountain Coal Field undertaken by your direction in the years 1891–92. The developments made show this to be a rich and valuable coal field. None of the seams have been fully explored, or tested, yet enough is now known of its contents to warrant the conclusion that this is destined to be one of the most productive Coal Fields of Alabama.

With grateful regards for past favors, this work is respectfully submitted.

A. M. Gibson.

East Lake, Ala., May 30, 1893.

REPORT.

To His Excellency,

GOVERNOR THOMAS G. JONES:

SIR:—I transmit herewith a Report upon the Coal Measures of the Blount Mountain Region, by A. M. Gibson, Assistant Geologist, who has devoted the greater part of the past two years to the examination of this region.

This report directs attention to a hitherto little known and less appreciated section of our State, and will, I feel sure, contribute materially to its future development.

I have the honor to be, Sir,

Your obedient servant,

EUGENE A. SMITH,

State Geologist.

UNIVERSITY OF ALABAMA,
July 1st, 1893.

Section I.

INTRODUCTORY.

The region lying between Murphree's Valley and Bristow's Cove on its N. W. side, and Cahaba, Coosa and Wills Valleys on its S. E. side is locally known by the name of *Blount Mountain.* It has long borne this name, rather to designate than to describe it. A part of it only can properly be considered a mountain, and only a part of it lies within the present, or the former confines of Blount county.

As the county boundaries now stand, the larger portion of it is in Blount, the balance in St. Clair and Etowah counties. The portions of it that may properly be classed as mountain, are its northwestern rim adjoining Murphree's Valley, and locally known as *Straight Mountain*—and its southeastern side which is still more elevated, and makes the top of the watershed between the Coosa and the Warrior drainage. This is the part that is properly called "*Blount Mountain.*" This is a very prominent elevation, extending the whole length of this coal field, and with more or less prominence extending north-eastwards to the Tennessee River. Its height in the region under consideration is generally from 1300 to 1500 feet above sea level.

. This coal area has been partially described in former publications, on "*The Warrior Coal Fields,*" and on "*The Plateau Region of Alabama.*" But these descriptions were necessarily brief and imperfect, a superficial reconnoisance only. They however showed this field to be one of importance—at least of sufficient importance to require more critical study and examination. The results of that study and examination are set forth as concisely as the subject would admit in the following pages.

An exhaustive report cannot be predicated on an untested coal field. Mining at different points, on the different seams, sufficiently extensive, can alone expose their qualities and geological characteristics. This only source of positive knowledge has been, and is yet, almost wholly wanting in this coal field. When the geological examination of this region was made no mining was being done ; the seams which had been opened in former years, were abandoned, and the openings filled up, or caved in. Some of these were re-opened, but others, and probably the most important, could not be, with the limited means at the disposal of the survey. Many new openings were made, and several seams of coal exposed, the existence of which had not been previously known. All of these will be referred to in their proper places in the details, and general description of the field.

GENERAL OUTLINES AND STRUCTURE.

This coal field extends from the S. E. corner of T. 14, R. 1 west in a north-easterly direction a distance of about forty miles, to the east side of T. 10, R. 5 east. From thence it gradually merges into "*The Plateau Region*" already described by Prof. McCalley In width it varies considerably, starting in with a breadth less than three miles at the south-western end, gradually widening out on the south east side till it attains a width of eight miles. At its widest part nearly opposite the town of Springville in St. Clair county, its southeastern edge is about three or three and a half miles from the northeastern end of the *Cahaba Coal Field*. It maintains its width of seven to eight miles to the middle of T. 13 of R. 3 east at Aughtery's Gap. From thence it gradually narrows around the edge of Greasy Cove. Opposite the middle of the Cove is its narrowest portion, scarcely three miles wide. From the township line between townships 11 and 12, R. 4 east, it widens perceptibly towards the northeast till it becomes 4 to 5 miles broad, which breadth it maintains to the upper or northeast end of the field. On its northwest side its margin is nearly a uniform straight line running nearly N. E. and S. W. It is the southeastern edge of Bristow's Cove and Murphree's Valley, to the lower, or southwestern end of the field.

Although the southeastern edge of the valley is the actual northwestern edge of the coal field, yet the high 'elevation called *Straight Mountain* composed wholly of coal measure rocks, intervenes between the valley and the productive portion of the field. This is a high, narrow ridge with vertical strata ; its general height is 300 to 400 feet above the valley. Its breadth from 1,000 to 1,500 feet, with 500 to 1,000 feet more of very highly inclined strata along its southeastern flank, or inner side. Hence, from 1,500 to 2,000 feet in breadth, along the edge of Murphree's Valley, which although it contains much coal, because not in available position for advantageous mining, must be regarded as unproductive. Toward the upper or northeastern end of Murphree's Valley this ridge gradually diminishes in height and volume, and northeast of the Locust Fork of the Warrior River its strata are changed gradually from the vertical to a highly inclined southeastern dip.

The dip of the strata in this field is mainly to the northwest, and generally about 10° to 12°, in a few places more, in many places less, except along the southeastern side of Straight Mountain which has a variable southeast dip, of 20₀ to 80°. Also a strip from one-quarter of a mile to two miles wide adjoining the mountain has a gentle southeast dip gradually diminishing to the horizontal. The field has the general features of an irregular synclinal with its lowest depression of strata near its northwestern side. The southeastern rim or *Blount Mountain* being elevated higher than *Straight Mountain*, the northwestern dips extend much farther than the southeastern; and these diverse dips do not meet as in a true or symmetrical synclinal fold, but both gradually diminish until they become imperceptible, and have a space of one to two miles broad of undisturbed horizontal strata between them. This general arrangement extends through the whole length of this coal field, but is more marked, and distinctly prominent in the wider portions. It may in descriptions hereafter be referred to as a *trough*, but it must be remembered that it is a trough with a horizontal bottom, and that bottom much nearer the northwestern than the southeastern side.

THE INTERIOR ANTICLINAL.

Through the widest portion of this field, there is a well defined *anticlinal* starting from near the southwestern end of Greasy Cove and running west by south. It is about ¼ of a mile broad ; it does not make a ridge, but is plainly and distinctly seen in the low lands where the streams have eroded the strata. It is particularly noticeable where it crosses Dearmond's Creek at the old Brasher Mill site, and Sand Creek at the upper end of the old Holt place. Also where it crosses Coal Bed Branch, and Difficulty Creek, to the Blackburn Fork of the Little Warrior River in T. 14, R. 1 east. Its course was traced about 10 miles, though it probably extends much farther. Its dip on the south by east and west by north sides is 10°, but its top appears to be unbroken. Its structure strongly suggests an underlying oil basin. And this suggestion is further strengthened by the fact that near the eastern end of this anticlinal, beneath the foot of the mountain a strong odor of escaping petroleum is perceived, and the slates of the lower coal measures are impregnated with oil, and burn freely in the fire. The underlying Trenton Limestone carrying much carbonate of magnesia is the great oil and gas producing rock—the source of the oil and gas in western Ohio, and eastern Indiana. This rock is of that character in the valley to the west, where it is largely exposed—whether it is of this necessary oil producing character beneath this anticlinal can only be determined by the drill. This matter must be left for future test and study.

At present we are only calling attention to the structure of this coal field, which has its symmetry in its widest portion marred and broken by this uplift passing diagonally through it.

DRAINAGE.

While the drainage from the top of Blount Mountain flows mainly with the dip of the strata to the northwest, into the head streams, or tributaries of the *Warrior River*; yet the area is distinctly divided into a series of basins by divides, or higher lands, which cross the field in a south-easterly and northwesterly direction, and make well defined watersheds between the head waters of these tributaries. The first of these divides near the southwest end of the field, crosses it near the surveyed line between Blount and St. Clair counties. Leaving *Straight Mountain* in section 19 and running east to section 22 in T. 14, R. 1 east is an elevation of land; a low flat divide between the *Canoe Creek*, and *Warrior River* drainage. From the top of *Buck Ridge* in said section 22, the dividing line of drainage is along the . ridge northeast for about one mile ; thence in a general northeasterly direction with the *Blount Mountain* to S. 8, T. 13 of R. 3 east. From thence in a general northwesterly direction along a high and broad plateau is the division of drainage between this basin, and the basin of the *Locust Fork* of the Warrior. The public road from Murphree's Valley to Whitney, generally, is near the top of this divide. from *Tait's Gap* in Straight Mountain, (S. 26, T. 12, R. 2 east) back to Heathcock's, and Aughtery's Gaps in Blount Mountain, in sections 16 and 21, T. 13 of R. 3 east. From Tait's Gap in Straight Mountain, southwestward, the boundary of drainage mainly coincides with that mountain to the beginning point. Only two small short streams break through it, one at the Allgood Gap, opposite the Champion Mines, the other at the Clowdus Gap three miles farther to the southwest. These two flow into the *Calvert Fork* of the Little Warrior. With the exception of these two small branches, all the drainage of the included area is into the *Blackburn Fork* of the *Little Warrior River*. This area may hence be appropriately referred to hereafter as the *Blackburn River Basin*.

SUB-DIVISIONS.

The Locust Fork Basin, begins at its southwest end on the high plateau of land which separates it from the Blackburn River Basin, thence to Blount Mountain on the southeast; thence with the top of said mountain to Gregory's Gap on the road from Walnut Grove to Attalla, in S. 24, T. 11, R. 4 east near the township line between ranges 4 and 5 east; thence north with said range line to Bristow's Cove. From Bristow's Cove southwestward with the edge of the cove, and the edge of Murphree's Valley to the line between Blount and Etowah counties; thence south along said line one mile to the top of the ridge north of the Hayse Gap; thence southwestwards along the top of this ridge to its junction with Straight Mountain near the south line of S. 18, T. 12 of R. 3 east; thence with said mountain to *Tait's Gap* the beginning point on the Blountsville & Whitney road.

' All this included area is drained by the *Locust Fork* of the Warrior River—and mav be 'properly called the *Locust Fork Basin.*

The upper, or northeastern end of this field, is a plateau region, drained mainly by *Line Creek,* and tributaries of Clear Creek, and other streams. It possesses a greater thickness of coal measures than the plateau region lying farther to the northeast, but not so great as the basins to the southwest. It carries several seams of coal that are identical with those in the Locust and Blackburn Rivers basins, and is hence an important portion of this coal field.

The remaining portion of the field, lying between the northwest boundary of the Locust Fork basin, and the Straight Mountain, is drained into the *Calvert Fork* of the Little Warrior River. It is a small area, but holds exposed some of the best, and most accessible coal seams in this field, and hence is of much economic importance.

AREA..

The area of this coal field is a little over one hundred and fifty square miles, exclusive of the sub-conglomerate measures, and of those that are vertical along, and near to Straight Mountain, and which for that reason may never be available or productive.

VERTICAL MEASURES OF STRAIGHT MOUNTAIN.

Much coal is shown to exist in the vertical and highly inclined measures of *Straight Mountain*, but its connection, · or identity with the workable seams of this field has not been found, and it is not believed ever can be. Only the leading, prominent rocks of this vertical formation are distinguishable. Of these, the first prominent hard rock on its northwest side is the *Lower Conglomerate*, and its underlying shales and rocks of the sub-conglomerate formation lie between it and Murphree's Valley, making the northwestern slope of the mountain. The next and most prominent rock near the centre of the mountain and standing vertical is the *Second Conglomerate*. This is the most prominent and conspicuous rock of this uplift. Towering far above all its associates in this compressed mass of vertical rock and strata—bold and naked, like an artificial wall, from twenty-five to one hundred feet above its congeners, like the caruncle, or comb of a rooster's head, and hence popularly called the *Cock's Comb*. (This is the great massive rock which makes *Buck Ridge*, and part of *Bee Ridge* hereafter to be described, and farther to the northeast, makes in many places the crest of Blount Mountain.) Between this rock and the *Lower Conglomerate* (which makes the falls on all the streams cutting through Straight Mountain,) is the place for the Caskie, Howard and Peacock seams of coal; but they have not been found, except in detached fragments, and pockets in the rocks. So great has been the contortion, and pressure, brought on this uplifted section, that all soft material, shale and coal seams, have been squeezed out, and all consecutive order obliterated, even in the intervening ledges of rock.

On the southeast side of the high vertical wall of *conglomerate rocks*, the rocks are compact, and vertical for one hundred and fifty to two hundred feet, then gradually assume a southeast dip with diminishing angles to the base of the mountain. A slight southeast dip continues for varying distances still farther to the southeast. Along the southeast side of the mountain, and especially near its base, seams of coal have been exposed, but all of irregular thickness, due evidently to greater or less compression, and also varying in strike from place to place. Some of this coal has been used in the shops, and with very satisfactory results; but all of these seams are too highly inclined to be mined to advantage; neither could any of these seams be certainly identified with those lying farther to the southeast. That they belong to that base group of productive measures lying above the *second comglomerate* is very evident, and probably near the middle of the group, and certainly *below* the *fourth conglomerate*, may be regarded as settled, but this is the closest approximation to their proper geological position as yet attained. The reason of all this will be apparent when the fact is considered that while this Straight Mountain is merely a fold of the coal measures, yet some portions of them are probably hidden by faulting, and that about 3,000 feet of strata are here compressed into a space of 700 to 1,000 feet horizontal measure, hence great compression, distortion, and displacement, necessarily resulted. These facts having been recognized, no further effort was made to develop, or study the *Straight Mountain* coals, and hence they are not included within the productive area of this coal field.

Also on the opposite, or southeastern side of this field is considerable area of lower or *sub-conglomerate coal measures*, which though they carry, at least in places, several seams of coal, yet all probably too thin to be of any practical value, and are hence not included here as productive measures. This area embraces all the coal measures which lie between the top of the sub-carboniferous, or *Mountain Limestone* and the top of the *Lower Conglomerate.*

SECTION II.

THE LOWER, OR SUB-CONGLOMERATE COAL MEASURES.

This belt or strip of measures varies in thickness from 600 to 800 feet. It is seldom that the top and base of it can be both seen at the same part of the mountain. Except in a few places the *Mountain Limestone* is wholly beneath the floor of the valley, and where it is much elevated the lower conglomerate on top has generally suffered much abrasion, and is in many places wholly gone from the crest, and the body of the rock, covered up by soil or clay can not be seen. As a close approximation to its breadth it may be said that this strip of *sub-conglomerate coal measures* usually occupies the steepest part of the mountain, and varies from half a mile to a mile in width, nearly the whole length of this coal field.

This is the formation so carefully and extensively described by Prof. McCalley, Assistant State Geologist, in his report on "THE PLATEAU REGION OF ALABAMA." This is the formation that makes the *"Plateau Region"* in the main, though generally reinforced by a thin mantle of carboniferous strata lying above, and concealing the *lower conglomerate.* Farther to the northeast in Alabama, Georgia and Tennessee, this formation frequently carries thick seams of excellent coal, but of uncertain extent, and varying thickness. Want of uniformity in thickness, and want of continuity in extent, of the coal seams are characteristics of this formation, suggesting the possibility that they were formed in *basins* of greater or less extent, and varying volume, and hence have not the uniformity of the coal seams of a later period of the carboniferous era.

It is generally believed that the coal seams of this forma-

tion thin out towards the southwest, and are not in the
Blount Mountain region of sufficient thickness to be of
value. This opinion may be correct, many known facts
seem to favor it, and yet it is an open question, the facts are
mainly of a negative character, and are not conclusive. The
same seam or seams occupying the same geological relations,
that are prominent, and highly productive, in Georgia and
Tennessee, have been traced from the Georgia line almost
to the southwestern end of Blount Mountain. Of course
this tracing was superficial and from the conditions on the
face of mountain could not be continuous, for the southeast
face of that steep mountain is generally a confused mass of
slides, and but little of its strata can be seen in place, especially where the coal seams belong. But wherever the
strata were found in place at the coal bearing horizon, and
not buried by slides or talus, there the seams, or clear evidence of coal seams were found. Fossil coal plants in great
numbers were seen in the rocks, almost continuously along
the face of the mountain, and these were generally referable
to the coal bearing horizons. This seems to be sufficient
evidence of the continuity of this as a coal bearing formation. And the fact that the seams which were seen, were
too thin to be worked, is not conclusive evidence, in this
formation where want of uniformity is the rule, that they
may not be thick in other places where they are not exposed.

Several prospect openings have in recent years been made
in the face of the mountain to find coal, some of them give
encouraging signs, but none of them have been driven in far
enough to make a fair test. It will require the expenditure
of much labor to make a satisfactory test at almost any
point, and while in the present state of knowledge assurance
of success cannot be given, yet neither should such efforts
be discouraged.

While these *sub-conglomerate measures* must in our present
state of knowledge, be classed as unproductive, yet they con-
stitute an important member in the State's geological col-

umn, and hence require careful description. They are more prominently exposed along the southeast edge of Blount Mountain and contiguous elevations to the east and north. east than anywhere else in the State, and thus come appropriately within the bounds of this report.

. These measures rest conformably on the top of the *Carboniferous* or *Mountain Limestone.* The transition, (where the junction was plainly seen,) is not abrupt, but the lime rock gradually passes into calcarneous shale, and the shale gradually shows more and more silica in its upper layers till it passes into flaggy sandstone, of dark grey color near the base, and gradually changes into light brown friable sandstone towards the top. Somewhere near the base of this member a thin seam of coal exists, its position was not found on the face of the mountain, but pieces of it were found washed out of a deep hole in Canoe Creek where it ran over the rocks of this member. As the creek in its course, had not run over any higher member of the coal series, it evidently belonged here. The thickness of this member approximates one hundred and thirty feet. It is succeeded in ascending order, by a few feet of soft shale and clay, and a seam of good coal one foot thick where seen. This is succeeded by dark slate—dark grey to black, soft and fissile, eighty feet thick, followed by flaggy sand rock for twenty or twenty-five feet.

The next member, which may be local, consists of two beds of variegated crystallized limestone or marble, the lower bed four feet thick, very hard and solid, holding many fossils, and a good deal of silica—a very beautiful strong rock on which the weather seemed to have made no impression. About 20 feet above this bed is another of similar material and color, but softer, having less silica and fewer fossils. This ledge is also solid and about 20 feet thick. The prevailing colors are a mingling of yellow. and brown. The thickness of this member where seen is about 45 feet.

This is followed by what is known as the *"Shaly Cliff,"*

a very prominent and persistent ledge of hard gnarly rock, often with bluffy perpendicular face 20 to 50 feet high. This ledge is the great impediment to road making across this mountain, it only can be surmounted at a few places. The whole thickness of this member is about 100 feet.

Following this is a series of flaggy and ripple-marked sandstones, yellowish and reddish colored building stone, with belts of intervening yellow clay, somewhat irregular in volume, the whole making from 50 to 75 feet.

Above this member and near the base of the conglomerate next above it is, at least in places, a seam of coal of varying thickness, but so far as seen not exceeding one foot thick. This coal seam occupies the geological position of the *Cliff Seam, Etna Seam*, &c., so often referred to and described in Prof. McCalley's Report on the "PLATEAU REGION OF ALABAMA" that no doubt of its identity is intertained.

The *Lower Conglomerate* caps and complete the series of *Lower Coal Measures*. It is a rock so well and widely known that minute description of it is unnecessary. It possesses here its usual characteristics ; a massive, persistent rock, varying in thickness from 50 to 100 feet, in some places, composed almost wholly of well rounded pebbles, firmly cemented together, and called *Millstone Grit*, in others the pebbles are in bands and patches, or sparsely scattered through the rock, while in other parts the pebbles are wholly wanting, and the rock is merely a coarse-grained sandstone.

The series of *Lower Coal Measures* will be presented at one view by the following :

SECTION OF LOWER OR SUB-CONGLOMERATE COAL MEASURES.

(11)	FIRST, OR LOWER CONGLOMERATE 50 to	100	feet.
(10)	COAL (*Cliff seam—Etna seam, &c.*)	1	"
(9)	*Flaggy sandstones* and clay beds, 50 to	75	"
(8)	*Shaly Cliff*,	100	"
(7)	Signs of coal seam? *shale*,	35	"

(6)	*Limy* or *Marble group*,	45 feet.
(5)	*Slate bed*—dark grey to black,	105 "
(4)	Coal—good quality,	1 "
(3)	Flaggy sandstone and soft brown-stone,	100? "
(2)	Coal—in Canoe Creek,	? "
(1)	Flaggy sandstone (estimated)	38 "
	Mountain Limestone.	

These measurements and estimates are probably too low in the aggregate; they were made at the lowest gaps, where the ascent of the mountain only was practicable. Besides this, while ascending the mountain towards the northwest, the dip of the strata, averaging about 10_n in the same direction, must be considered, this would carry the base of the measures a good deal lower beneath the top of the mountain than at its base. Hence it is considered that about 200 feet must be added to the measurements and estimates of the foregoing *section*, thus bringing up the whole vertical thickness to about 800 feet. In other words that from the average top of the mountain would require a perpendicular measure of 800 feet to reach the *Mountain Limestone.*

SECTION III.

THE UPPER OR TRUE COAL MEASURES.

Under this head are embraced all the known productive measures in this coal field. All that lie above the *First*, or *Lower Conglomerate.* The designation of this rock as the *First Conglomerate* is preferred, because there are *four distinct conglomerates* in this field, and it is easier, and more natural to designate and refer to them by their consecutive numbers, and natural order, than by any other means. These conglomerates are the guide lines, the geological land marks in the study of a coal area. They separate each series of coal seams from all others, and enable each to be examined in detail with greater certainty. Hence, the study of the conglomerate rocks of this field is first presented.

THE SECOND CONGLOMERATE.

This is the rock so often referred to in Prof. McCalley's report on the "PLATEAU REGION OF ALABAMA" and elsewhere, as the "*Upper Conglomerate.*" It is truly designated there as the "*upper,*" because it is so in fact, there being no other conglomerate above it. It is there generally the top rock of the series. But in this field it lies comparatively near the *base* of the *productive coal measures.* Probably the greatest horizontal distance between the top of the *First Conglomerate,* and the *base* of the *Second,* is 3050 feet, while the thickness of strata intervening is 336 feet. These figures are probably too high for a general average of thickness of strata, or distance between these rocks; but they were obtained at the only place found where measurements could be made with approximate accuracy; and are fairly representative for a large portion of this field.

This rock resembles the *First Conglomerate* in many respects, and general structure, yet differs from it so much in general appearance, that the one need never be mistaken for the other. Its pebbles are generally smaller, less crystalline, and less firmly cemented together, and where pebbles are wanting the rock is generally of a lighter grey color, and of more coarse quartzitic structure than the other. It is also in the region under consideration, much more massive and prominent than the First Conglomerate, often rising up very boldly above the surface with its outcrop covering a space of from a quarter to three quarters of a mile wide. That the thickness varies greatly is evident, but at no place has it been found possible to measure it, except by the space it occupies.

In the lower, or southwest end of the field, it comes near to the surface over a wide space, making all that portion drained into *Canoe Creek,* a plateau region. Its southeastern outcrop makes the ridge known as *Buck Ridge* from Fall Branch in T. 14, R. 1 E., and running thence north eastwardly through that Tp. and T. 14, R. 2 E., and T. 13, R. 3 E. It

crosses the Tp. line between T. 13 and T. 12 in S. 4 of 13
and 33 of T. 12, R. 3 E.; and in S. 34, T. 12, R. 3 E., near
Walker's Gap, it reaches to the top of the mountain; and
with the exception of two short spaces it makes the main
top to S. 13, T. 12, R. 3 E., where it towers in the lofty
naked masses known as *"Buzzard Rocks."* Opposite this
point it was found that the *First Conglomerate* made the
lowest bench on the southeastern face of the mountain, and
that the Lower Coal measures were mainly buried beneath
the floor of *Greasy Cove*, and that the coal seams lying between the *First* and *Second* conglomerates have their out-
crops on the southeastern face of the mountain. Opposite
here is the narrowest part of this coal field, the part lying
between Greasy Cove, and the headwaters of the Locust
Fork of the Warrior River; the cause of it is obvious, the
uplift which produced Greasy Cove encroached largely on
this coal field—indeed was wholly formed out of it; leaving
only a margin of two and a half miles between its rim, and
the Murphree's valley fold.

From the Gilland Gap, near the "Buzzard Rocks," in
S. 13, T. 12, R. 3 E., the *Second Conglomerate* continues to
make the top of the mountain to Tumbling Gap in S. 6, T. 12,
R. 4, E. At this gap this rock is much broken up and dis-
placed. The heaviest boulders are on the southeastern side
af the mountain, and the First conglomerate still farther
down towards the foot than at the Gilland Gap. This position of these rocks continues to the northeastward to Gregory's Gap in S. 19, T. 11, R. 5 E. on the road from Walnut Grove to Attalla; there the outcrop of Second conglomerate is wholly on the southeastern side of the mountain.
It is only 100 feet thick, and is separated from the First conglomerate by only 110 feet of strata, while at Lyttleton on
the T. & C. R. Road, near Line Creek in Secs. 17 and 20, T. 11,
R. 5 E., it is at the level of the valley, and only separated from
the vertical First conglomerate by a line of fault a mere
seam in the rocks. But this is abnormal; for after passing
up Line Creek above *"Buck's Pocket"* it assumes its normal
2

position on the face of the bluffs; and at the Sheffield Gap
in S. 9, T. 11 of R. 5 E., it is again found at the top of the
mountain.

Crossing the coal field northwestwards to *Bristow's Cove*
the *Second Conglomerate* is again found making the upper
part of the bluff on the east side, and extends with the bluff
toward the southwest for several miles. This rim of the
cove gradually declines in height toward the lower end of
the cove, and sinks beneath the surface before reaching the
Locust Fork of the Warrior River, and this rock too passes
beneath the surface and is not seen again till it rises up in
the middle of Straight Mountain near the Tp. line between
Tps. 11 and 12 of R. 3, E. From thence it continues prominent to the southwestern end of Straight Mountain, as a
vertical rock.

The Third Conglomerate

Is not a prominent rock, and may not be co-extensive with
the coal field. Its place is near the middle of the productive measures. It is a coarse dark-colored rock in its upper
parts, and near the base a reddish conglomerate formed of
good-sized, but not well-rounded pebbles, firmly cemented
together with carbonate of iron. Its place can be better ex-
plained when it is reached in the description of the coal
seams.

The Fourth Conglomerate

Is among the upper members of the coal measures in this
field. There are but two seams of coal lying above it. It
occupies but a small portion of the field, and is only found
in a high flat-topped ridge lying between *Straight Mountain*
and *Ponch Creek*, and the *Warrior River* below the mouth of
Ponch Creek. It underlies the most of sections 17, 18 and
19 and part of 8, 9 and 3 of T. 12, R. 3 east.

The rock in the upper part is light-colored, loosely
cemented, weathers badly, and is hence seldom seen on the
surface; but its place is plainly shown by a profusion of

well-rounded, large-sized pebbles. Its lower part is irony, harder and better preserved. Owing to its disintegration and consequent want of face exposures, the thickness of the upper part could not be ascertained, but it is evidently thin —probably 10 to 15 feet thick. The lower or irony portion of this conglomerate is seemingly thicker. At one place its outcrop measured 40 feet, that might have been greater than the average.

For 100 feet below this conglomerate bed, the rocks are all quartzitic, many of them very granular. These, with the conglomerate above, to which they are allied, making a stratum 150 feet thick, and containing several good seams of coal, are peculiar to this section of the field, and make the top member of the Blount Mountain Coal Measures.

Whether this top series of conglomerate and quartzitic measures is the equivalent of the "*Montevallo Conglomerates*," the top series of the CAHABA COAL FIELD, or not, has not been determined, but is rendered somewhat probable by the geographical relation of the Fields, and their general similarity of structure. That one is the continuation of the other is evident. Both belong to that great strip of coal area cut off to the southeast by the great Jones Valley fold, and running a southwesterly direction; the direction in which it is known the Alabama Coal Measures mainly thicken. These Fields were originally *one*, and are now only separated by the branch valley, three miles wide, which connects the Jones Valley with the great Coosa Fold. It was, therefore, reasonable to expect a greater similarity between the strata and coals of *this Field* and the CAHABA immediately to the southeast of it, than between it and the WARRIOR FIELD farther to the northwest. This expectation has been reasonably verified in the progress of this survey.

The other prominent rocks and features of this Coal Field will be noticed under the head of "*Details.*"

EXPLORATION OF THE FIELD.

In the beginning of the investigation of this Field a persistent effort was made to first develop the vertical and

highly inclined coal seams, in and near the base of Straight
Mountain, and to assign them to their true position and relation to the other seams. Examination of seams was made
and measurements taken in the *Cowden Gap* in section 28, in
the *Waide Gap* in section 27, and in the *Clowdus Gap* in
section 13, all in T. 13 of R. 1 east. Also in the *Allgood Gap*
in S. 4, T. 12 of R. 2 east, with the following results : The
outcrop of the *Cowden Seam* is in the S. E. of S. E. of S. 28,
515 feet southeast of the vertical rocks, average dip of strata
in this space 15° southeast, height of seam above the highest of the vertical rocks, at the southeast base of Straight
Mountain, 171 feet.

This seam varies in thickness from 3 feet 9 inches to 4
feet 6 inches. The coal is of low grade, impure, cokes imperfectly, gives a red ash. It shows here the following

SECTION:

Roof, yellowish-brown shale.................Seen 6 feet.
COAL.......................................30 inches.
Slate, dark, soft...........................4 inches.
COAL.......................................16 inches.
   Bed Rock, hard, gnarly, yellow-gray sandrock.

No other outcrop of coal was found up the branch toward
the southeast from which the relation of this seam to those
above or below it could be inferred.

About a mile southwest of the Cowden Gap opening, on
"Dry Branch," a seam of coal was exposed which was
classed as identical with the Cowden Seam, which showed
the following

SECTION:

Roof, slate...
COAL, slaty................................30 inches.
Shale, parting.............................7 inches.
COAL.......................................15 inches.
Clay, dark................................4 inches.
   Bed Rock, hard sandstone.

Location in N. W. of S. W. of S. 33, T. 13, Range 1 east. This coal is of a little better quality than the seam in the Cowden Gap.

On the north side of the Blackburn Warrior in N. W. of N. W. of S. 4, T. 14, R. 1 east, another opening apparently on the same seam gave this

SECTION:

Roof, dark slate...
COAL, fair...................................30 inches.
Shale, parting................................3 inches.
COAL, good..................................12 inches.
    Bed Rock, sandstone.

In the *Waide Gap* in the N. E. of S. E. of S. 27, T. 13, R. 1 east is a seam of coal which has been worked, and tested for some years past to a small extent, and known as the *Waide Seam*. This seam crops out 1,000 feet southeast of the vertical rocks of Straight Mountain. Its opening shows a slight northwest dip, which is doubtless produced by a small local wave. Coal of better quality than at the openings previously mentioned. It shows the following

SECTION:

Cap Rock, hard, curly sandstone.................10 feet.
Slate, nearly black............................25 feet.
COAL..................................2 feet 2 inches.
Slate, soft....................................2 inches.
COAL...............................1 foot 4 inches.
Under Clay, dark.............................4 inches.
    Bed Rock, dark sand-stone.

This coal gives a red ash, cokes reasonably well, is hard and glossy, with mainly cubical fracture. It has generally been regarded as identical with the Cowden Seam, yet as many points of difference, as of agreement with that seam, are shown in its structure and surrounding. Its true position is not free from doubt.

In the *Clowdus Gap* in S. 13 the structure is yet more complicated. From the northwestern side of the Lower Conglomerate, measured southeastward, at right angle with the strike, to the first known vertical coal seam, the distance is 545 feet, and thence to southeastern side of vertical rocks 450 feet, making whole thickness of vertical strata 995 feet. The highly inclined strata dipping S. E. gives the following

SECTION:

(1) From southeast side of vertical rocks to first known coal seam..............................50 feet.
(2) COAL, thin and irregular...........3 to 12 inches.
(3) Slate, dark blue...................32 feet.
(4) COAL, thin seam...................3 to 4 inches.
(5) Slate, light blue to grey.................300 feet.
(6) COAL, good, bright, hard, cokes well......1 to 3 feet.
(7) *Slate*, gray, and hard brownish gray sandstone.50 feet.
(8) COAL, hard, brittle, cubical, (Saw Mill seam).22 inches.
(9) Slate, hard light blue to gray..............60 feet.

It is evident from inspection of this section, that no point of identity with similarly located sections farther to the southwest, either in coal seams or strata, can be clearly perceived. It may be that seam No. (6) is the equivalent of the Waide, or Cowden Seam; but it carries much better coal, and is crushed and distorted to such a degree that its identity could only be a matter of inference.

The bed-rock of seam No. (8) alone, of all the strata seen in this section, bears a strong resemblance to the bed-rock of the Cowden Seam. But in all other particulars the two seams are wholly dissimilar.

That a fault, or slip in the strata, of unknown extent, exists in this gap, is evident from seams above No. (6) of the last section, not bearing the same relation to it, or to each

other, on both sides of the gap.   One-fourth of a mile to the
northeast of the line of the last section in southeast quarter
of S. 18, T. 13, R. 2 east, we find Nos. 7, 8 and 9 of that section are wholly wanting, and are replaced by the following
above No. 6:

FAULT SECTION.

(6)   COAL, good, bright, hard, cokes well......1 to 3 feet.
      Shale, hard, irony.........................20 feet.
      COAL, soft, thin seam.....................4 inches.
      Shale, hard, and sand-rock thin.............40 feet.
      COAL, blackband seam..............2 feet 2 inches.
      Sand rock, reddish, soft, and clay to top of
         hill...................................20 feet.

Here, at least 110 feet of former section are replaced by
totally different strata, that probably belong to a higher
level.

The upper or Black Band seam shows the following

SECTION:

Roof hard, reddish sand-rock.....................6 feet.
COAL..4 inches.
Clay, parting..................................2 inches.
COAL..4 inches.
Black Band..................................4 inches.
Clay, parting ....2 inches.
COAL..8 inches.
Fire Clay.......................................

On the southwest side of the Gap (Clowdus) unconformity
of strata is equally as apparent, but being more involved
in the uplift of the Mountain, a fair section of it could not be
obtained.   The highest seam there found with a southeast
dip was the Saw Mill Seam, No. 8, of the Gap Section.   Its
outcrop was found in a gully where there was much water,
and it was so highly inclined that no satisfactory test could
be made on it.   Its thickness and importance are yet un-

known. Where its outcrop,was first seen in the Gap, on nearly level ground, it was only 10 inches thick. A test pit was sunk 15 feet farther to the south, and in that distance it had increased to 22 inches of solid coal. But the seam being under water, and dipping downwards 8° to the southeast, no farther test could be made at that place. Should it continue to increase in thickness as indicated it will prove a valuable seam of coal.

From this point all the seams that were known, or could be found, were cut in succession across the Field, at its widest part, or toward the Arch. Walker Gap, in S. 36, T. 13, range 2 east. All the streams running into the Blackburn Fork of the Little Warrior, from both sides, were carefully searched for coal exposures, or indications, and all such found were tested. Then measurements were made across the Field from the top of the Lower Conglomerate at the eastern top of Blount Mountain, as nearly at right angles with the strike as practicable, to ascertain the thickness of strata between the respective seams, and the aggregate thickness of the coal measures.

To present the data thus obtained in a concise form, and to give greater clearness to the general description of the seams, a general section of the whole Field is given, including the upper seams, found only in the basin of the Locust Fork of the Warrior.

GENERAL SECTION OF THE SUPER-CONGLOMERATE MEASRURES.

Top of Measures—Clay, irony shale, soft sandstone.................................20 feet.
(28) COAL, soft, no cap rock....................2 feet.
Shale, irony clay, sand-rock, some pebble....30 feet.
(27) COAL, the *Bynum Seam*............4 feet 4 inches.
The 4th conglomerate and quartzite rocks...80 feet.
(26) COAL, seam not named; good, bright outcrop.........................1 foot 8 inches.
Coarse sand-stone, flaggy.................40 feet.

(25) COAL, *Carnes, Paine, Smith, Gaither*
seam.........................3 feet 8 inches.
Quartzitic, sand-stone and clay............60 feet.
(24) COAL, *Baine seam. Phillips seam?*...3 feet 4 inches.
Flaggy sandstones, slates.................65 feet.
(23) COAL, *Fossil slate seam*............1 foot 2 inches.
Soft sandrock shales and clay, about......100 feet.
(22) COAL, *Woodward seam*......2 feet to 3 feet 3 inches.
Hard arenaceous shale, and flaggy sand-
rocks............................40 to 60 feet.
(21) COAL, *Armstrong seam*.....3 feet to 3 feet 5 inches.
Thin flaggy sandrock and slates..........87 feet.
(20) COAL, Thin seam—*Guide, or fossilliferous
seam*....................................1 foot.
Clay slate mainly.......................44 feet.
(19) COAL, Thin seam not named.............8 inches.
Sand rock 15 feet, flaggy rock and slates
20 feet...............................35 feet.
(18) COAL, Thin seam.
Heavy bedded sand rock 20 feet, shales
and slates 60 feet......................80 feet.
(17) COAL, Thin seam.
Heavy sand rock 10 feet, flaggy sand rock
25 feet, shaly slate 25 feet, clay slate
12 feet...............................72 feet.
(16) COAL, Thin seam, no name.
Heavy bedded sand rock 20 feet, flaggy
sand rocks 80 feet, clay slate and clay
2 feet...............................102 feet.
(15) COAL, Thin seam, *The Farley*.............1 foot.
Rock, hard massive, 25 feet, thin sandy
slates 90 feet, clay slates 26 feet......141 feet.
(14) COAL, Double seam, separated by fossiliferous shale, The *Sally hole* seam, *Mace
Murphree* seam ....1 to 2 feet.
Clay, slate and flaggy sand rock...........69 feet.
(13) COAL, The *Adkins seam, Clements seam,*
etc.............................4 feet 6 inches.

Flaggy sand rock........................81 feet.

(12) COAL, Thin seam—*Adkins spring seam.*
Flaggy sand rock increasing in thickness....82 feet.

(11) COAL, The *Jourdan seam*...........2 feet 3 inches.
Flaggy rock, heavy bedded sand rock,
slate.................................57 feet.

(10) COAL, The *Ivy hole seam.* Third Conglomerate ....1 foot 7½ inches.
Shaly and thin bedded sandstone, slates....55 feet.

(9) COAL, The *Murray seam, Washington seam,
Jackson seam*...........................3 feet.
Soft shales, clay slates, hard gnarly rock
and dark slates.......................100 feet.

(8) COAL, Found only in deep water; supposed *Waide seam*...............3 feet 6 inches.
Dark slates; sand rock, flaggy irregular....53 feet.

(7) COAL, *Big seam—Holt seam.* Outcrop in
river drilled through; has several partings; whole thickness..........12 feet 6 inches.
Thin sandrock and gray sandy slates, by
aneroid measure.......................80 feet.

(6) COAL, *Sand Creek, No. 2.* Thin seam........1 foot.
Heavy hard sand rock, coarse flaggy
slates.................................129 feet.

(5) COAL, Soft impure in bed of Sand Creek.
Sand Creek, No. 1 ....2 feet.
Hard compact rock, generally massive.....312 feet.

(4) COAL, *Garner, Lowe, Hullett, Aderholt,
Brasher, Harris seam*....................2 feet.
Flaggy to compact sandstone 250 to 300
feet, Conglomerate, 2nd 400 to 495 feet..795 feet.

(3) COAL, *Howard seam*................3 feet 6 inches.
Prismatic shale and clay slate..............125 feet.

(2) COAL, *Caskie Seam*................3 feet 8 inches.
Reddish colored sandstone, hard shale,
and compact building stone.............336 feet.

(1) *Peacock Seam*—at outcrop.........1 foot 7½ inches.

Sand rock of various grades to top of 1st
or Lower Conglomerate..................56 feet.

Whole thickness of upper measures.......... 3,415 feet.
And of lower measures (see section)........:.... 800 "
Making whole depth of this Coal Field....... 4,215 "

SECTION IV.

DESCRIPTION OF COAL SEAMS.

DETAILS.

Seam No. 1, called the *Peacock Seam* from the iridescent
sheen, or luster of part of its coal, was not discovered till
near the close of the work in this field. It hence was not
as fully tested and traced out as could have been desired.
It was first discovered by digging a small pit on the lands
of John C. Martin in S. 8, T. 14, R. 2 east, and in sinking a
well in the southeast corner of said section; the seam was
dug through near its outcrop. It was there 19 inches thick,
solid coal, of very fine quality. This was 12 feet below the
outer edge of the cap rock, and was not there considered a
fair test of the thickness of this seam. It is believed from
all the observed indications that this seam, when fully
tested, will be found to be of workable thickness, and an
important seam of coal. Indications of this seam could be
traced southwest for about half a mile only, from the place
where it was discovered. The rocks with which it is associated gradually thin out, and were not found any further in
that direction.

Toward the northeast plain evidences of it were found as
far as the Wm. Walker Gap in S. 34, T. 12, R. 3 east. It
may not extend the whole length of the field. Probably it
will be found only in that portion where the strata between
the *First and Second Conglomerates* are thickest in townships
12, 13 and 14 of ranges 2 and 3 east. Over this part of the

field the space between these two conglomerates exceeds
500 feet, while toward either end it is diminished to 100
feet or less.

The composition of this coal, as determined by Dr. J. M.
Pickel of the University of Alabama, is as follows :

ANALYSIS OF COAL OF PEACOCK SEAM.

Moisture............................ 1.49
Volatile, combustible matter.......... 32.38
Fixed carbon....................... 61.46
Ash 4.67
                                    _____
                                    100.00

Coke 66.13
Sulphur 1.79

The *Caskie Seam* No. 2 is apparently one of the best coal
seams in this field. It was first opened by a Mr. Caskie
many years ago, in the northeast corner of S. 10, T. 14, R. 2
east. An entry was driven in about 20 feet, and the coal
coked and otherwise tested with very satisfactory results.
This opening showed the following

SECTION :

Hard prismatic shale...........................20 feet.
Blue roofing slate.............................. 3 feet.
COAL, with some thin clay partings.........3 feet 8 inches.
Under clay, dark, sandy.........................2 feet.

There is good evidence that this seam extends toward the
southwest as far as the divide between the waters of the
Blackburn Fork and Canoe Creek; and to the northeast as
far as the examination of this field was extended, though
probably not generally carrying its normal thickness.

At Maldin's Gap, in the S. W. ¼ of S. E. ¼ of S. 27, T. 12,
R. 3 east, the *First and Second Conglomerates* are both
plainly seen about a quarter of a mile apart. This distance,

with a dip of 10° northwest, would show the thickness of strata between them to be less than 250 feet. About the place where the Caskie Seam would be looked for there are good signs of coal. One large Chalybeate spring probably comes from this coal seam. Seemingly the coal is near the surface, but it would be an unfavorable place to prospect or dig for it. The Gap is a low one, with a cross fault running through it; much water accumulates, and many springs break out here which would be a great impediment to mining. This is the gap through which the T. A. & C. R. R. Co. have run the line of survey for the location of their road.

From this gap toward the northeast the second conglomerate makes the top the mountain mainly, and the seams lying beneath it are found on the face of the mountain. The debris and slides on the steep slope towards Greasy Cove have generally obscured the outcrop of this seam for many miles. Its position was seen and recognized at Gregory's Gap, in the southwest corner of S. 19, T. 11, R. 5 east, and also in the same section on one of the branches of Clear Creek. In this region the seam is thin, and the strata between the first and second conglomerates is reduced to 110 feet by aneroid measurement.

On Line Creek, in sections 8 and 17, T. 11, R. 5 east, this seam has again become thicker. Several openings have been made on it by Dr. Dozier and others. The works were closed up and a section could not be obtained. The reputed thickness is 30 inches solid coal. Its position was also seen here on the lands of Mr. Gordon, on the north side of the creek.

This seam was next found in S. 1, T. 11, R. 4 east, where it had been opened by G. B. Waide. The opening was made on the eastern rim of Bristow's Cove, 90 feet above the floor of the cove. The first conglomerate, not seen, its place is beneath the surface. The dip of the seam here is 60° east at the surface, but it is gradually lessened after passing the immediate rim of the cove to 10° or less in 200 yards. The opening was caved and filled up, but from Mr. Waide was obtained the following

SECTION:

Hard arenaceous sand rock.....................10 feet.
Blue clay, or decomposed blue slate...............4 feet.
COAL, bright, glossy, nearly solid.........3 feet 10 inches.
Clay, parting.................................4 inches.
COAL, cubical.................................4 inches.
    Under clay.

This coal, so far as seen, appears to be of good quality, and is reported to coke well and to work well in the blacksmith forge. It has not been as yet otherwise tested.

From the regularity of the uplift in the southeastern edge of the Cove, it is probable that this seam may be found at the proper level both above and below the Waide opening for a space of eight or ten miles. Toward the Locust Fork of the Warrior River it evidently is sunk downwards, passes beneath the river and is not seen any more on the northwestern side of this field.

The *Howard Seam* No. 3 of the general section is found at the widest part of the field 125 feet above the Caskie Seam, and just beneath the base of the great second conglomerate rock. An opening was first made on this seam by Mr. Howard on the S. E. of N. E. of S. 9, T. 14, R. 2 east. It shows the following

SECTION:

Hard massive sand-rock at face...................12 feet.
Yellowish-gray shale and blue slate..............10 feet.
COAL, hard, bony................................1 foot.
Dark shale parting....................½ inch to 2 inches.
COAL, good.............................2 feet 6 inches.
Fire Clay, bluish-white, very fine.................2 feet.

This seam, like the former one, (No. 2,) will not probably be found farther to the southwest than the drainage of the

Blackburn Fork of the Warrior extends in that direction. From thence toward the northeast its place is easily recognized, though often deeply hidden by debris piled up over its outcrop at the base of the second conglomerate. This base rock, which forms the cap rock of this seam, is generally a coarse sand-rock, seldom carrying pebbles, but always of such coarse grain and light gray color as to be readily recognized as a member of the second conglomerate.

At Malden's Gap, in S. W. of S. E. of S. 27, T. 12, R. 3 east, it is about a quarter of a mile northeast of the top of the mountain. Between Malden's and Gilland's Gaps it is on the very top of the mountain. It had here been dug into by Mr. Bynum years ago and some coal exposed, but not fully developed. In the region around Gilland's Gap, in N. E. of S. W. of S. 13, T. 12, R. 3 east, its place is about 200 feet below (S. E. of) the top of the mountain. A bold chalybeate spring breaks out here, known as the Gilland spring. Some digging had been done here and some coal and blackband ore exposed. The rocks contain many fossil coal plants and the position of the seam here is easily determined.

At Tumbling Gap in southeast of S. 6, T. 12, R. 4 east, its place was found on the southeastern face of the mountain, about 200 feet from the top. The strata here are too much broken up to warrant the successful opening of the seam at this place, though much more favorable places can be found on either side.

In S. 19, T. 11, R. 5 east, this seam had been opened by Mr. Copeland in two places. The seam is thin here, so far as cut into, only 15 inches thick, though its maximum certainly had not been reached. The coal is of fair quality, and further development may show better results.

On the breaks of Line Creek in sections 8 and 17, T. 11, R. 4 east, the strata are much broken and contorted, yet plain evidences of this seam were observed at several places. Near the second railroad bridge, on the north side of the creek, is a mural face of rock believed to be the second conglomerate, 125 feet thick, beneath which there is evidently a seam of coal; coaly, or bituminous shale, 4 feet thick, is exposed here. This is believed to be the outcrop of the Howard seam. Also, in section 9, same township, near the railroad track, a seam 20 inches thick is exposed in the railroad cut, which is provisionally considered to be the same. On the southeastern edge of Bristow's Cove the evidence of this seam was observed about 60 feet above the *Caskie seam*, and beneath the coarse disintegrated conglomerate which makes the top of the bluff, or eastern rim of the cove. It has not been opened anywhere on this side of the field, but though closer to the Caskie seam than on the other side of the field, no doubt is entertained of the identity of this and the Howard seam.

Like the other seams lying between the first and second conglomerates, this seam is not found on the northwestern side of the field at any place southwest of the Locust Fork of the Warrior River.

The *Lowe, Hullett, Garner seam*, No. 4, of the *General Section*, is the first known seam above the *Second Conglomerate*. It has long been known and highly esteemed for its uniformly excellent quality of coal. It is a very persistent seam, and bears throughout an almost uniform thickness of 23 inches. The quality of its coal very closely resembles the *Black Creek* seam of the Warrior Coal Field, but it does not occupy the same geological position. In a few places it has been seen only 21 inches thick, in others 26, and more rarely 30 inches. Over most of this field this is a solid seam of coal, but toward the southwestern end it has a small clay parting near the top. Mr. D. Aderholt has had some mining done by tunneling on this seam, in the W. ½ of S. 4, T. 14, R. 2 east, and his opening gives the following

SECTION:

Flaggy reddish colored sand rock..................15 feet.
Heavy shale.......................................6 feet.
Blue slaty clay ....5 feet.

COAL, hard bright cubical ....9 inches.
Clay parting..1 inch.
COAL, hard lustrous...........................14 inches.
Fire clay bluish gray..............................2 feet.

In parts of this tunnel there were 27 inches of coal, but it again diminished to 23 inches.

An opening in S. 17, T. 14, R. 2 east, gave nearly the same section—coal 20 inches.

Two openings on the same seam on the east side of S. 11, T. 14, R. 1 east, gave this section:

Roof: shale......................................4 feet.
COAL, good.'....................................8 inches.
Shale, parting..................................1 inch.
COAL, good....................................12 inches.
Fire clay, floor.

This seam was not opened or traced further toward the southwest, but it probably extends in that direction as far, at least, as the divide between the Blackburn Warrior, and Canoe Creek drainage.

In a northeast direction from the Aderholt mine this seam is found on the east fork of Difficulty Creek, a little above its junction with the west fork, and is known as the *Hullett Seam.* On Coal Bed brarch, in S. W. ¼ of S. 34, T. 13, R. 2 east, it was long ago opened and known as the *Lowe Seam.* On the N. E. ¼ of same section, on the west prong of *Sand Creek* it is called the *Garner Seam.* And in the N. E. of S. W. of S. 26, same township, at the Old Brasher Mill seat on *Dearmon Creek*, it was exposed at the side of the creek and known as the *Brasher Seam.* It is not seen and has not been opened on the divide between the Blackburn river basin and the Locust Fork basin, but is again found in the latter on Hurricane Creek, below Buttermilk Falls, near the junction of the east and west branches of the creek in S. W. ¼ of S. 32, T. 12, R. 3 east. .Its outcrop here showed 23 inches of solid coal, with surroundings similar to the sections already given. Farther to the northeast its outcrop is

3

on the east side of Bee Ridge, and crossing the upper
branches of the Locust Fork and branches of Pole Creek,
about a mile northwest of the top of the second conglomerate.   It was also opened at Harris' coal bed, in N. E. of S.
E. of S. 15, T. 12, R. 3 east.

COAL SEAM No.—.   On Hurricane Creek where it passes
through *Bee Ridge*, in southwest of S. 32, T. 12, R. 3 east,
near to the out-crop of Coal Seam No. 4, and 30 to 40 feet
higher, is a Coal Seam, which is not given in the GENERAL
SECTION, because it was not found elsewhere, though reported
to exist also on Dearmon Creek, at a similar height above
the *Brasher Seam*.

What the thickness, or importance of this seam may be,
are matters of conjecture ; its out-crop is beneath a heavy
impending bluff which would crush it far below the normal
thickness at the edge of the seam, even if it were not otherwise reduced by weathering.   It shows this

SECTION.

Solid reddish cap rock, flaggy in upper part.... 20 ft.
COAL, good................................. 0 ft. 6 in.
Shale parting........................... .... 0 ft. 2 in.
COAL, good................................. 0 ft. 6 in.
Fire clay.................................... 1 ft. 6 in.
Clay Slate, with nodules of clay ironstone...... 6 ft.

Over this seam there is no covering of slate, but the coal
is in immediate contact with the massive cap rock.   It is a
little wavy, and irregular in thickness, but this probably
results from the unequal pressure of the overlying bluff.

That there may be other seams existing between Coal
Seams Nos. 3 and 4 is very probable.   They are over one
mile apart, surface measure, and separated by nearly 800 feet
of vertical strata.   That all this mass of measures should be
destitute of coal is improbable, but as yet no evidences of
any have been discovered in this space.

COAL SEAM No. 5* is also known as the *Sand Creek* Seam
No. 1. It does not give promise of being a seam of much
importance. It was cut on the bottom lands, and in the bed
of Sand Creek in S. 27, T. 13, R. 2 east, and on the hills
facing the eastern Branch of Sand Creek, coal soft and
brittle, seam about 2 feet thick, not used. It was also found
in southeast of northeast of S. 31, T. 12, R. 3 east, in the
Locust Fork Basin. Its position is about 315 feet above the
Lowe, Garner, Seam No. 4, of the General Section.

About 130 feet above the Sand Creek seam No. 1 is found
the *Sand Creek Seam* No. 2, No. 6 of the *General Section*.
It carries good coal, but is a thin, and consequently unimportant seam, cut in southeast corner of S. 21, T. 13, R. 2 east.

Eighty feet above Sand Creek Seam No. 2 is the *Holt*, or
Big Seam, No. 7 of the *General Section*. The out-crop of this
seam was found only in the bed of the Blackburn Fork, and in
deep water in Dearmon Creek. It could not be tested, or
exposed by digging above water level, nor could samples of
its coal be obtained for analysis. Testing was done by drilling through the seam, the record of which gave the following detailed

SECTION.

Roof, clay slate............................	5 feet.
COAL......................................	0 foot 4 in.
Clay Slate.................................	1 foot 5 in.
COAL......................................	0 foot 3 in.
Clay	1 foot 5 in.
COAL......................................	1 foot 6½ in.
Rock parting..............................	2 feet 0 in.
Slate and *Coal*...........................	1 foot 3 in.
Coal and Clay.............................	0 foot 6 in.
COAL, hard................................	1 foot 6½ in.

* No. 5 is called in the vicinity, the *Upper* Sand Creek seam, and
No. 6 the *Lower* Sand Creek seam, but as these names give a false
idea of their relative position, we have substituted the names Sand
Creek Seams No. 1 and No. 2.

Clay . . . . 0 foot 6 in.
Blackband and *Coal* . . . . : . . . . 1 foot 11 in.
Clay . . . . 0 foot 4 in.

Hard rock supposed to be bed rock.

Whole thickness of seam 12 feet 6 inches.   Drilling done in southwest of southwest of S. 22, T. 13 of R. 2 east.

It cannot yet be known if this drilling gives an average section of this seam.   The small amount of under clay here, seems not to be in proportion to the thickness of the seam. And the absence of any clay above the rock parting excludes the idea of its being a *double* seam.   Then the rock parting may be local; at one opening made on the Howard seam No. 3, a very hard rock was encountered imbedded in the coal; that opening was abandoned, and another made 100 yards farther east, where no such rock existed.

Where this seam crosses Dearmon Creek in N. E. ¼ of S. 22, T. 13, R. 2 east, a better view of its surroundings was obtained ; though the seam being cut by the creek in deep water nothing·could be there learned of its size or structure. Large amounts of coal were washed out here by the great flood of July, 1872.   Wagon loads of it were gathered up and hauled off in after years.

About 55 feet above this seam is another one cut by this creek, No. 8 of the *General Section.*   It also is cut in deep water, and while the coal could be distinctly felt with a pole, yet only, the surroundings of the seam could be seen.   It is overlaid by 8 to 10 feet of dark slate, and 10 feet of hard gnarly cap rock.   This rock weathers out rough and scaly, giving it an appearance readily recognized.   The slates and this cap rock very much resemble, if they are not identical with, the slates and cap rock, over the *Waide seam* in S. 27, T. 13, R. 1 east, heretofore mentioned.   While this evidence alone is not conclusive on the question of identy of seams, yet taken in connection with the fact that no other seam in this Field has as much similarity in surroundings, gives their supposed identity a degree of probability.   From the evidence obtained it seems to be a necessary inference that

the *Waide seam* on the northwest side of the Field must be the equivalent, either of this seam, or of the one next above it, the *Murray seam*, next to be described.

This seam has not its out-crops exposed farther to the northeast than the middle of S. 23, T. 13, R. 2 east; or if so, it has not been observed. This may arise from the fact that the dip of the seam. and the steep slope of the hills, are both in the same direction, and all signs of it are hidden by the descending debris. Towards the southwestern end of the Field there are two places where coal is torn up in deep holes in the river; one in S. 31, T. 13, R. 2 east; the other in S. 10, T. 14, R. 1 east, which from their location are believed to be on the southeastern out-crop of this seam.

No. 9 of the *General Section* is known as the *Murray seam*, because it was first discovered on the lands of John Murray in northeast of northwest of S. 22, T. 13, R. 2 east. It was here cut by the river, and exposed in its bed, mostly under water. Coal, bronze shade, brittle breaking up into small cubes, good quality, cokes well, a good blacksmith coal. As exposed here the seam was only 17 inches thick, in two benches, with clay parting of 4 inches near the middle. Believing that a larger seam of coal existed here than was shown in the river, and to test it, and to show its relations test pits were sunk on the seam 200 yards to the southeast and one-quarter of a mile to the southeast of the river, and also a drill hole sunk to bed rock below the seam in the river. These tests showed that there were two small subsidiary seams of coal, one 8 feet *above*, the other nearly 4 feet *below* this seam, and that all the intermediate slates were a *mass of fossil coal plants*. The coal seam was just the same in the test pits as where cut by the river. It was made evident, however, that enough carbonaceous matter existed here to make a large seam of coal, if circumstances elsewhere admitted of its being all combined into one bed. That this should be the case in some other part of the field is very probable, and gives this seam a great prospective importance. It shows at this place the following

SECTION:

Cap rock, hard, rough, compact, shaly.....10 feet
Shales and clay slates....................6 feet
Coal, poor...............................0 foot 3 inches
Slates dark, very fossiliferous.............8 feet
COAL, brittle, cubical..........................7 inches
Clay parting.................................4 inches
COAL, brittle, cubical.........................6 inches
Clay................................. ......2 feet 0 inches
Slate, fossiliferous.......................1 foot 8 inches
COAL ....6 inches
Under clay..................................4 inches
    Bed rock.

The cap rock of this seam being easily distinguished from
other rocks, was traced in an east to northeast direction to
where it crosses the Blountsville & Asheville road, about
three-quarters of a mile northeast of Foster's Old Chapel,
thence north with the slope of the hills to the river again at
the lower end of P. Clements' place, in S. W. of N. W. of S.
23, T. 13, R. 2 E., very near the corner of the section; may
be in the N. E. of N. E. of S. 22. The seam was found here
to be much improved, and the slates above it to have but
few fossils. Coal very greatly improved. The upper bench
of it has long been known, and highly appreciated by the
local smiths. It is classed by them as the best coal in the
Field, some of large experience say it is the best coal they
ever used. The lower and more important part of the seam
was discovered and brought to light by the Geological Survey. It is a notable feature of this field, and also of the
Cahaba, that many of the seams have clay partings; these
partings have in many instances been mistaken for *under
clay*, and the most important portion of the seams overlooked. Other instances of the same kind will be given
hereafter.

This seam shows here the following

Section:

Slate in river bed..............................5½ inches
Coal, bright, glossy, good....................10½ inches
Clay parting..................................7 inches
Coal, like upper bench.......................18½ inches
Under clay.....................................2 feet
Hard slate.............................1 foot 8 inches
   Bed rock.

In one mile this coal has increased in thickness 16 inches. It is here in a bed 29 inches thick, exclusive of under clay and roofing slate. Ample thickness for mining, and carrying coal that would take the lead in any market.

The seam could be opened near this place on either side of the river, above water level, and with natural drainage.

Whether this seam continues to increase toward the northeast cannot now be known, as its outcrop is covered by the bottom lands of the Blackburn Warrior in that direction to the broad divide between that stream and the Locust Fork basin. That it can be found and opened in this space there is no doubt, but it will require careful search to locate it.

Northeast of the divide this seam was certainly seen at but one place in the Locust Fork basin. On a branch on the lands of John Jackson, near the northeast corner of S. 32, T. 12, R. 3 E., the peculiar cap rock of this seam was observed. Coal had here been worked out of the upper bench of this seam 14 to 18 inches thick. The clay parting, as in many other places, was doubtless considered the base of the seam. If the lower bench, which no doubt exists here as elsewhere, has thickened in proportion with the upper one, or even to the same extent, this is here a very fine seam of coal. A proportionate increase of the lower bench of coal which is reasonable to expect, would give here over 4 feet of coal.

Toward the southwestern end of the field this seam is certainly known to have been cut at but one place, near the mouth of Armstrong's Creek in S. 31, T. 13, R. 2 East. It is there known as the *Washington bed*. Coal has been taken out here in considerable quantities in former years, though the works are now filled up.

The coal here, as in all other places on this seam, is in two benches, with a clay parting between. Coal of very fine quality about 20 inches thick. Not thick enough for mining, here, or probably anywhere else in the southwestern part of this field.

Immediately above this seam at the Washington beds is the *third conglomerate*. Generally it is an irony conglomerate holding rather large-sized pebbles, and very firmly cemented together with carbonate of iron. It may not be co-extensive with the field; was only seen at, and south of, the Washington bed, and on the river bluff at this horizon, in S. 21, T. 13, R. 2 East.

No 10, of the *General Section*, is about 55 feet above the Murray Seam, and is locally known as the "*Ivy Hole*" seam, so called from the large amount of ivy growing near to the deep hole in the river where this seam was first observed.

At the place where it was first opened near the southeast corner of the N. E. of N. W. of S. 22, T. 13, R. 2 East, it was only 13½ inches thick, and so far as traced on the river bluffs to the southwest seemed about to hold its own. But to the northeast it thickens a little. In the S. E. of S. E. of S. 15, T. 13, R. 2 E., it was drilled through at the edge of the water and found to be 19½ inches thick. Should this increase continue a mile farther northeast it would become a workable seam.

From 55 to 60 feet higher up—strata, flaggy sandrock—is the Jourdan Seam, No. 11 of the *General Section*. It is called the "*Jourdan Seam*" because it was first opened on Jourdan Creek in the N. W. ¼ of S. 14, T. 13 of R. 2 East. This was a very promising outcrop of coal, capped by a heavy ledge of sand rock, and lying above water level.

Its outcrop showed the following

SECTION:

Cap rock, heavy bedded sand rocks..........8 feet
Slate and shale...........................2 feet
Coal....................................7 inches
Slate parting..............................3 inches
Coal ....1 foot 5 inches
   Fire clay.

It was expected that this would develop into a workable seam. The coal was evidently of good quality, but further development did not show any increase in the thickness of the seam; and at other openings made on it farther to the southwest the seam was thinner and less promising than on Jourdan Creek. It may be increased in volume towards the northeast, and in the divide between the Blackburn and Locust Fork basins, but in the latter it was found to be a thin seam.

Eighty feet above the Jourdan Seam is the Adkins Spring Seam No. 12 of the *General Section*. It is a thin and unimportant seam. The intervening strata generally above and below it are thin bedded, flaggy sandstone, with intercalated beds of clay slate.

Eighty feet above No. 12 is the *Adkins Seam* No. 13 of the *General Section*.

This seam was partly exposed in the bed of the Blackburn Fork of the Little Warrior. It had long been known, and its coal was highly appreciated. But all the coal taken out here for years past was out of the upper bench of the seam, which alone was known to exist. An examination made here by the Geological Survey showed that the clay beneath this coal had not the characteristics of *under clay*, but of a clay parting in a coal seam. The drill was applied and the lower and most important portion of the seam discovered. Location in N. E. of N. W. of S. 22, T. 13 South, R. 2 East. This seam gives the following

Section:

Roof, clay slate.

Coal, good.................................9½ inches
Clay parting.............................1 foot 6 inches
Coal....................................2 feet 2 inches
Under clay...............................2 feet

Whole thickness of coal nearly 3 feet.
Thickness of seam, 4 feet 5¼ inches.
Dip of strata, 4° N. W.

It was readily seen that a short distance below, or south of this bed, this seam could be opened above water level, and at a right angle with the strike, so as to secure natural drainage. A most eligible and desirable position for mining. This coal is known to be an excellent shop coal—that determines its merits, because a good shop coal must necessarily be a good coking and furnace coal. It must be low in sulphur, high in fixed carbon, coking well and easily, and producing but little clinker; hence a good shop coal must necessarily be a good all-round heating, steam and furnace coal.

It is a very peculiar fact that the *Murray Seam* No. 9 and the *Adkins Seam* No. 13, lying within less than 300 feet of each other, and carrying the best coal as yet passed over in this description, have both very thick clay partings, the removal of which will probably somewhat increase the cost of mining the coal, but the coal from either seam, when put on the market, will command a price that will fully compensate for any increased cost of its production. Both seams are so situated as to afford excellent facilities for operating self draining mines above water level, and accessible to any mode of transportation.

The next 70 feet of strata above the Adkins, or Adkinson Seam No. 13, are mainly composed of thin, smooth sandstones, with occasional intervening bands of hard, sandy slates.

Above these is coal seam No. 14 of the *General Section.*
It is known as the *Mace-Murphree Seam,* the *Baldwin Branch
Seam* and the "*Sally Hole*" *Seam.*

This seam was first opened on the lands of Mace Murphree, in the N. E. of N. E. of S. 12, T. 13, R. 2 East, but
more fully tested on both sides of the Blackburn Fork, and
also at a deep hole in said fork, known as the "Sally Hole,"
all in section 15 of the same township. Also examined
where it crosses Jourdan Creek, one mile farther east in
section 14.

This seam is of but little value wherever it has yet been
opened. It consists of a thin seam, or stratum, of coal at
the top, then an immense bed of slate filled with fossil coal
plants, with a small well-defined seam of coal, and under
clay at the base.

There is more carbonaceous matter scattered through the
great included mass of slates than would have been required
to form a thick seam of coal. In this respect it resembles
the Murray Seam No. 9, where that was first opened; but
differs from that in having a much greater thickness of included fossiliferous slates. A clear idea of its structure will
be obtained from the following

Section:

Cap rock, heavy, solid	10 feet to 15 feet
Slate roof	4 feet to 6 feet
COAL	3 inches to 1 foot
Fossiliferous slate	20 feet to 30 feet
COAL	9 inches to 1 foot
Under clay	2 feet
Bed rock.	

This seam is very peculiar in the large space it occupies
—upwards of 30 feet from *bed rock* to *cap rock.* No other
seam, except the Murray, has been found to approximate it,
anywhere, at least in this field. Its existing conditions were
evidently produced by the infiltration of sedimentary matter
among the coal or peat-making plants, during the ages of

their growth, which prevented their being consolidated into one body. That the same conditions, for an indefinite period of time, should have extended over the whole of this coal field, would seem to be exceedingly improbable. Hence the expectation may be reasonably indulged, that somewhere in this field, not yet discovered, the ample carbonaceous matter of this seam may be found in one large compact coal seam.

The great thickening of the Murray Seam, under like conditions, towards the northeast somewhat strengthens and confirms this expectation.

The *Farley Seam* No. 15 of the GENERAL SECTION lies about 140 feet above the big fossiliferous seam. Strata mainly thin slaty sandrock. Coal seam opened many years ago on the Farley branch. Coal good, but too thin to be mined.

There is probably another seam lying between the two last mentioned, but if so, it is also a thin one.

Between seams Nos. 14 to 20 there are about 500 feet of strata of very similar structure, mainly thin slaty sandstone, and beds of thin sandy slate, and holding many thin seams of coal. Each of these seams is capped by a harder and more compact ledge of sandstone. These ledges of cap rock very plainly mark the position of the several seams in this thin seam belt. Neither of these six seams was anywhere found to be thick enough for mining, and they are hence regarded as of but little value.

Probably the best one of this thin seam series is the upper one, No. 20 of the *General Section*. It carries throughout very good to excellent coal, and has been much sought after for blacksmiths' use. In the low grounds and along the streams in sections 16, 17, 18 and 20 of T. 13 R. 2 E. it has been obtained by stripping off the thin overlying strata. The seam here is seldom over a foot thick, often less, and overlaid by 6 to 8 feet of very fossiliferous shale. The fossils in this slate are numerous, distinct, and often of large size. They mark the horizon of the seam so well that it is readily located by the exposed fossils. It was hence called the *Guide seam*, because it was a well known landmark, and aided in assigning other seams to their proper horizons. This seam is lost sight of in the divide between the two main basins of this field, but makes its appearance again in the Locust Fork basin. A good deal of coal has been taken from it in sections 19 and 20 of T. 12, Range 3 E. The coal is equally as good as in the Blackburn River basin, and a little thicker—12 to 14 inches thick. The overlying fossil-filled shale is replaced by about the same thickness of fossil-filled dark *fissile slate*. This is the only seam that is known to carry this kind of slate. It mines up in large square blocks, like roofing slate, and splits smoothly one way. Wherever split there is at least one beautiful impression of a fern leaf, or frond, on one or both sides of the slate.

This seam was also seen on the bottom lands of Dry Creek in S. 19, T. 11, of R. 4 E. about three miles east of Walnut Grove. Slate fissile, and full of fossilleaf-impressions. This coal seam was stripped and raised from the bed by *Warren Haynes*, occasionally, for many years. It is not worked now, but is widely known as the *Haynes bed*. This seam was only seen at these two places in the Locust Fork basin. The position of the Haynes bed is close to the elevated southeastern margin of Bristow's Cove—the continuation and representative of Straight Mountain—with the trend of the seam closing in toward that uplift.

This condition is due to the fact that above, or northeast of, the Locust Fork of the Warrior River, the trend of the Cove is *east* of northeast, and that in the course of about twelve miles, this eastward trend has cut into the Blount Mountain Coal Field about one and a half miles. Hence all of the coal seams near its northwestern side are cut by the edge of the cove at an acute angle, and end or terminate in that uplift, which makes the southeastern rim of the cove. Hence this seam cannot be found any farther toward the ·northeast. This opening was made on its southeastern outcrop, and at or near its most northeastern extension.

This coal was analyzed by Dr. J. M. Pickel, of the University of Alabama, with the following result:

ANALYSIS OF COAL—HAYNES SEAM.

Moisture......................	1.27
Volatile combustible matter...........	36.49
Fixed carbon.......................	56.19
Ash	6.05
	100.00

Coke	62.24
Sulphur	3.87

Above the Guide Seam No. 20 the strata are generally clay slate and thin sandstone; in some places wholly clay slate; in others, mainly hard slate and sandstone for fifty to sixty feet, to seam 21.

The *Armstrong Seam,* No. 21, of the *General Section,* was first opened on Armstrong's Creek, in S. 17, T. 13, R. 2 E., by W. B. Armstrong, about sixty feet above seam 20. The cap-rock over this seam is very massive, but not in all places solid rock, Generally it is a rock that weathers into a rough scaly mass, giving it the appearance of compacted shale. In other places the rock is solid and smooth in its lower portion, and only scaly and scragly in its upper parts, yet always presenting a type, or idiosyncrasy of structure that could be very readily recognized, and the seam traced without difficulty. It extends throughout the upper half of the Blackburn Fork basin, and is found in the Locust Fork basin, and probably extends through the divide between them. Towards the western end of the Blackburn Fork basin, this seam either becomes small and insignificant or is divided into two seams. Near the western side of S. 20, T. 13, R. 2 E., the characteristic cap-rock of this seam is found about 40 feet above Seam No. No, 20, with a thin six inch seam of poor coal beneath it. This is at least 20 feet below

its normal position, and more than 80 feet below the Woodward
Seam next above it, yet no evidence of an intervening seam
was found between them, where the upper member of the
Armstrong Seam properly belonged. It is probable that
this seam either divides or terminates near this place. It
was not traced or recognized southwest of the range line
between ranges 1 and 2 east.

To the northeast of S. 20, T. 13 of R. 2 E., its position is
very plainly indicated by its very prominent cap-rock,
through the remainder of this basin till it passes beneath
the divide between this and the Locust Fork basin. Its
cap-rock is again seen in S. 20, T. 12, R. 3 E., near the Locust
Fork, and finally on the edge of Dry Creek, in S. 19, T. 11,
R. 4 E., at the very edge of the uplift that makes the rim
of Bristow's Cove. It is here about 40 feet above the guide
seam, or *Warren Haynes'* Seam last described. It may be
of workable thickness here, though its outcrop showed only
18 inches of coal.

Where first opened in the N. E. ¼ of S. 17, T. 13, R. 2
E., this seam presented the following

SECTION:

Cap-rock, hard, solid, shaly on surface............15 feet.
Roof, bluish slate, and reddish shale...............4 feet.
COAL..4 inches.
Slate and black band.........................6 inches.
COAL..7 inches.
Clay..7 inches.
Coal..8 inches.
Slate and pyrite..............................2 inches.
Coal..6 inches.
Under clay....................................——————.

Whole thickness of seam here 40 inches, exclusive of roof-
ing-slate, and under clay—ample thickness for mining, but
there were too many partings, and the character of the coal
was not satisfactory.

Another opening was made in this seam in the S. W. ¼ of the same section which gave better results—seam thinner at out-crop, but carrying much better coal. This opening gave the following

SECTION:

Cap-rock solid, upper part shaly.................15 feet.
Roof, light-blue slate.............................4 feet.
COAL, good, bright..............................3 inches.
Slate and blackband...........................6 inches.
COAL, good......................................6 inches.
Clay, dark-gray................................7 inches.
COAL, good, clean.............................8 inches.
Slate...1½ inches.
COAL...2½ inches.
Under clay....................................—— ——.

While this opening gave but 34 inches at the edge of the cap-rock, yet the seam thickened one inch per foot as far as the entrance was made. This increase in thickness gave assurance that the seam was thick enough for easy mining; dip south by east 4 deg.; above water-level, self-draining. The coal was found to be a good shop coal, coking well, and holding but little sulphur.

Its analysis by Dr. J. M. Pickel is as follows:

ANALYSIS OF COAL—ARMSTRONG SEAM.

Moisture............................ 1.46
Volatile combustible matter.......... 31.11
Fixed carbon...................... 64.97
Ash............................... 1.46
                                  ———
                                 100.00

Coke............................. 66.43
Sulphur.......................... 3.87

This is an excellent seam of coal of the very best quality and giving an exceedingly small percentage of ash.

Above the cap-rock of the Armstrong Seam the strata are hard arenaceous shale and hard flaggy sand rocks, increasing in hardness and thickness to the bed rock of the Woodward Seam, which is a solid, compact, yellowish-gray sand rock 2 to 3 feet thick.

The whole thickness of intervening strata, between the Armstrong and Woodward seams is forty to sixty feet.

The Woodward Seam No. 22 of the *General Section* has been cut in many places in sections 18 and 19 of T. 13, R. 2 E. An opening was made on this seam in the S. W. of the N. E. of s—d S. 19 with the following results: Seam penetrated about four feet, when work was stopped by wet weather, and the caving in of the cut. The seam at the end of the cut presented the following

SECTION:

Roof, soft slate.

Coal, cubical.....................................9 inches
Slate parting, probably local......................½ inch
Coal, good.......................................10 inches
Clay parting.....................................3 inches
Coal, bright, hard..............................16¼ inches

Under clay.

Whole thickness of seam 39 inches; the seam thickening and the quality of the coal improving as the work progressed. A sample of the coal last taken out upon analysis by Dr. J. M. Pickel yielded the following result:

ANALYSIS OF COAL—WOODWARD SEAM.

Moisture............................ 1.17
Volatile combustible matter........... 34.90
Fixed carbon....................... 59.52
Ash 4.41
                                    ———
                                    100.00

Coke 63.93
Sulphur............................ 2.20

4

This seam as observed at many other places is always in two benches. It has always a clay parting near the middle of the seam. The small slate parting in the upper bench in section 19 was not seen at any other place. It was hence inferred that it was only a local peculiarity, and would not continue far. The seam may therefore be considered as a two-bench seam, carrying 16 to 18 inches of coal each, and separated by 3 to 4 inches of clay, which is easily removed and will much facilitate the operation of mining and working this seam.

This coal has a fine reputation as a good shop coal. It cokes well and easily, is low in sulphur, and possesses a large amount of fixed carbon, giving it free combustion and endurance in the furnace.

At some places where this seam has been opened the upper bench has been found too bony for good blacksmith coal, and better suited for grate coal, yet this peculiarity is also local and does not obtain universally. Where this seam was opened by the Survey there was very little difference in the grade or quality of the coal in the two benches of the seam, the upper one being just a little more splintery, and mining out in larger blocks than the lower one was the only observable difference.

This is doubtless a fine seam of coal, thick enough for mining, and occupying a position in the field high above water level, and lying throughout its extent almost horizontal, will give all desirable facilities for easy mining.

The Woodward Seam is the upper one over a large extent of this field. It may be regarded as the top seam of the Blackburn River basin, so far as it extends. It is found on all the high lands to the north and northwest of the Blackburn Fork. It may extend through the divide between the two basins, but has not been seen or recognized in the Locust Fork basin. It probably exists there also, but is thinner and less prominent than in the Blackburn River basin.

Above the Woodward Seam the strata are mainly fine ar-

gillaceous shale of varying thickness, according to the elevation of the country, from 40 to 125 feet.

These strata are easily eroded and have been much gul-
lied and excavated, and now present a surface of level plateau flanked by steep-sided ravines and gorges, and long, dry
hollows and low ridges.

Near the divide between the two basins there has been
less erosion, and the country is in the main an undulating
plain, only broken by the small streams which make the
headwaters of the two forks of the Warrior River. On this
high plateau there is no well marked top of the water shed,
but streams are found interlocked and flowing in opposite
or diverse directions into the Calvert Fork, the Blackburn
Fork and the Locust Fork of the Warrior River.

On the northwestern side of the Field, near the top of the
divide in S. 26, T. 12, R. 2 East, near the Tait Gap, in
Straight Mountain, the surface becomes more elevated towards the northeast, and a higher series of coal measures
exists. These upper measures increase in thickness and
elevation for two miles in a northeasterly direction, attain-
ing their maximum elevation in sections 17 and 19 of T. 12,
R. 3 E., but continuing to, and across, the Locust Fork of
the Warrior River.

In this space it must also be noticed that the measures
decline or sink towards the basin of the Locust Fork, and
hence the upper measures have a greater thickness here
than is apparent from the elevation of the country. It has
not been found practicable to get an accurate measurement
of the amount of this declination of the measures towards
the Locust Fork. Yet it is very perceptible on both sides
of the river. The depression appears to centre, or be the
greatest, at, or near, where the river passes out of this Coal
Field. This roll or depression may not much exceed a hundred feet. It cannot be less than that and certainly cannot
exceed two hundred feet.

But it is enough to put the lowest seam of the upper
measures below water level at its southeastern outcrop, its

highest point. *Water level*, however, is always a relative term, dependent on the natural drainage of the locality. Yet the base of these upper measures here is so near the level of the river that it is apparent that they are absolutely lower by from one to two hundred feet than they were a few miles to the southwest.

The lowest seam of this upper series of measures is No. 24 of the *General Section*, a very fine seam of excellent coal. It was opened near its southeastern outcrop, on the headwaters of Pouch Creek, by Mr. Phillips, and is known as the *Phillips Coal Bed.* The opening is in S. 19, T. 12, R. 3 East, about 50 yards east of the range line, and within a 100 yards of the southwest corner of the section.

The coal was reached by stripping off the surface clays, three to six feet thick. No cap rock or roofing slates above the coal at this point. Coal of good quality, in one solid bed, 3 feet 2 inches thick. The works were filled up when examined and a section of the bed could not be obtained. The dip of this seam at this place is 10° to southwest, this is doubtless due to some local roll in the measures. A quarter of a mile to the northwest, where the rocks are well exposed, the dip is very regular, about 8° to northwest.

This seam of coal could be drained and mined at this place, and when cut into beneath a solid roof would doubtless yield a still better quality of coal, with a probably increased thickness of seam. A short distance from the opening the edge of the cap-rock of this seam crops out, but its thickness could not be ascertained.

The rocks above and around this seam are all quartzitic, composed of distinct and well rounded grains of quartz. All of them show the saccharine or sugar-loaf texture; while some have the fish-roe, or öolitic texture. This peculiar lithological structure extends from a little below this seam, with slight variations to the very top of the measures. Hence this upper group of coal seams may be properly designated as the quartzitic, or upper conglomerate group.

This is a very interesting and valuable group of coal

seams. It does not cover a large extent of country, but holds throughout its extent *three* very fine seams of coal, besides several smaller ones in a vertical thickness of one hundred and eighty-five feet.

The only other opening beside the Phillips yet made on the lower seam of this group, is about a mile north of the Phillips bed, and on the opposite side of a high ridge in the S. E. of N. E. of S. 18, T. 12, R. 3 E. This opening was made many years ago, and coal taken out from time to time by many different parties. The opening is locally known as the *Lower Baine bed.* The opening was made by sinking a pit in a flat hollow, near the side of a branch, and below water level.

Many statements, probably exaggerated, were made about the thickness of this seam by the parties who had dug coal here. They generally agreed that "the seam was over 6 feet thick." Others said "the seam was standing on edge, and was more than 6 feet across." These statements were made by men of average intelligence and unquestioned veracity.

To settle the uncertainties about the size, structure and identity of this seam, it was determined to clear the pit of water if possible, so that the seam could be seen and examined. With the aid of several hands, and after long and persistent labor, this was accomplished.

The seam was found to occupy a normal position—dip 6 deg. east, and gave about the following

SECTION:

Blue shaly clay and thin coal seams........2 feet
COAL, solid, face and butt structure.........3 feet 4 inches
Clay, apparently under clay.

It is not absolutely certain that the clay beneath this coal is *under clay*; it may be a clay parting in the seam. The inflow of water was too great to admit of testing any deeper. It can only be positively stated that it had the appearance of *under clay.*

This coal is of very fine quality, and the seam is evidently one of great value. Though the seam where opened is below the water level, except in very dry weather, yet there is fall enough that could be used, without heavy expense, to make a large portion of this seam self-draining.

An average sample of this coal analyzed by Dr. J. M. Pickel gave the following results:

ANALYSIS OF COAL—BAINE BED.

Moisture	1.18
Volatile combustible matter	32.35
Fixed carbon	64.18
Ash	2.29
	100.00
Coke	66.47
Sulphur	0.92

This is a very remarkable seam of coal—practically it carries no sulphur. It will be largely utilized in the future for smelting iron and furnace work. That this opening and that of the Phillips bed are both on the same seam was very satisfactorily settled by their relative dips, their equal distance above the slate beds, and their environment by the same peculiar class of rocks. All these points of agreement left no room to doubt their identity. The Baine opening shows two inches more coal, and of better quality than the Phillips bed, but these differences probably arise from its better protection; and at neither place is the seam well enough protected to have preserved its full normal thickness, or its best coal.

This seam, lying as it does at the base of the quartzitic group, is necessarily more extensive than those above it. And though it has not been opened at any other place, yet this is probably due altogether to the fact that its position has not heretofore been recognized, and the seam searched for. Lying as it does near the base of the ridges, with its

out-crop often hidden by the debris of the higher lands, it
is no wonder that it has been overlooked, especially since it
was in a position where no seam of coal was expected to be
found or known to exist.

It may be confidently looked for above the slates and near
the base of the quartzites, on both sides of the high ridge
which rises between the head waters of Ponch Creek and
the Drury Bynum Creek in S. 19, T. 12, R. 3 E., and extending to the Locust Fork of the Warrior, and also for a mile
or more northeast of that stream.

Its position may also be approximately found from its relation to the well known *Carnes, Paine* and *Smith* seam,
which lies about 60 feet above it, and is well marked by a
heavy, massive cap-rock that generally shows its position.

The *Carnes, Smith, Paine* seam, No. 25 of the GENERAL
SECTION.

This seam is better known than any other in this coal
field. It is the only one on which coal mining to supply the
market has successfully been carried on. A tunnel was
driven in on this seam in S. E. of S. 8, T. 12, R. 3 E. by G.
B. Carnes, who supplied the local demand for coal for several years. Since then other openings have been made on
the seam by *Smith, Gaither, Paine* and others. The coal is
uniformly good, coking well and working well in the forge.
We have no analysis of the coal of this seam, but in all
practical tests to which it has been subjected it has given
satisfaction.

The best exposure of this seam that was examined is in
the S. W. of N. E. of S. 18, T. 12, R. 3 E. It there presented
about the following

SECTION:

Shale and thin soft sandstones.......... 10 feet
Cap rock hard, quartzitic, wavy........3 to 8 feet
Blue slate roof......................3 to 4 feet
COAL, hard, bright, cubical............. 3 feet 8 inches
Under clay, fine, dark................. 3 feet
Dip of strata, 25 degrees to southeast.

This high dip evidently does not continue far in that direction. In the S. E. of N. E. of same section, about one-fourth of a mile east from this opening, and at several other places the southeast dip is 3 degrees to 5 degrees. It probably diminishes to about that declination within a hundred yards to the southeast of this opening.

In the southwest ¼ of section 8, same township, an opening was made on this seam by Mr. Samuel Smith, but a tunnel was not driven in on the seam. This opening exposed the following

SECTION:

Cap rock, coarse-grained sand rock.................15 feet
Roof, reddish clay slate.......................... 8 feet
COAL, good, nearly............................... 4 feet
    Under clay.
Dip southeast, 3 degrees.

CARNES' COAL MINE.

Half a mile farther to the east in southeast ¼ of the same section, is the *Carnes* opening and tunnel, on the same seam. *Coal*, 3 feet 8 inches, solid, good quality. Cap rock and roof similar to the above section. By aneroid measure the Carnes mine is 75 feet higher than the Smith opening, but the dip of the strata is here 5 deg. west, showing a local elevation or roll in the strata. The direction of the dip will fully account for the difference in elevation, while the similarity of the coal and all the enclosing strata fully proved the identity of the coal seams at the two places.

THE GAITHER OPENING.

An opening on this seam has been made by *Mr. Gaither* in the northeast corner of S. 10, T. 12, R. 3 East. Coal covered up when visited said to be 4 feet thick. A wide opening had been made here, and the bluff cut into the solid rock, but no tunneling done. The coal is said to be of very good quality, but when visited could not be seen, being buried by

slides from the top of the cut.　Dip of strata here 5 deg. to
northwest.

About half a mile to the northwest of the Gaither opening, and on the northwest side of the same ridge, Mr. Gaither
had cut a coal seam 3 feet thick which lies 20 feet lower by
aneroid measure.

It was supposed that both openings were upon the same
seam of coal. There is much room to doubt the correctness
of this supposition; the dip of 5 degrees would, in half a mile,
carry the seam first opened, much more than 20 feet below
its level at the opening in section ten (10), and the associated
rocks at the second opening do not add anything to sustain
the claim of identity.　It is probable that the second opening here on the N. W. ¼ of S. E. ¼ of S. 4, T. 12, R. 3 East is
on a coal seam not yet named or identified which lies between this seam and the Bynum Seam, and which is num-
bered 26 in the General Section.

The *Carnes-Gaither* Seam No. 25 has also been opened
and mined by *James Smith* on the bluff of the Locust Fork,
near the southeast corner of S. 2 in T. 12, R. 3 E.　The coal
has a fine reputation, and there is a ready home market for
all that has yet been mined here.　When visited, mining had
been for some time suspended and the drifts so blocked up
that measurements of the seam could not be made.　The
seam is well known to carry good coal throughout, to be
solid, and upwards of three feet thick.

By aneroid measurement it was found that the seam here
is 70 feet lower than at the Gaither opening in the northeast
corner of section 10, showing the downward flexure of the
measures towards the Locust Fork of the Warrior, heretofore noticed.

Northwest from the James Smith mine on the opposite,
or northwestern side of the same ridge, this seam was also
opened on the lands of *Mr. Zach. Paine*, near the line between sections 2 and 3 of the same township.　Coal same as
at the other openings.　Seam between three and four feet
thick.

This seam has also been cut by James Smith on the hills
or river bluff, northeast of the Locust Fork, but the prospect was not so good as on the southwest side of the river.
Probably the seam becomes thinner northeast of the river.
The hill or high ridge which contains this seam does not
extend far in that direction. One mile northeast of the
river it becomes broken up into knobs, and these soon become less and less prominent, until they sink into an undulating plateau, which does not contain any of the coal seams
of the quartzitic group.

The downward flexure of the coal measures, in this basin,
has its greatest depression at, or near, the Locust Fork of
the Warrior, in a southeast and northwest direction; and
northeast of the river the measures are gradually elevated
till they attain their normal level. The lower coal seams of
this group must necessarily come to the surface within one
or two miles northeast of the river. The measures here
contain much less rock and hard strata than exists southwest of the river, and are more abraded. The outcroppings
of the seams have not been discovered; there are no rocky
ledges to designate their positions, and it is very probable
that they very materially thin out towards the northeastern
end.

Number 26 of the *General Section* is a seam that as yet
has received no distinctive name, and about which but little
is known.

It lies immediate between the *Upper Baine, Carnes Seam*
No. 25, and the *Bynum Seam* No. 27 of the General Section.

Some prospector first cut this seam in a pit on the north
side of the long high ridge which holds the quartzitic group
of coal seams, in the S. E. of N. E. of S. 18, T. 12, R. 3 E.
Its position is about 40 feet above the *Upper Baine, Carnes*
seam, and probably about the same distance below the *Bynum Seam*. The pit was not cut far enough into the hill to
reach or expose any cap rock, or probably to show the full
thickness of the seam.

The roof was shale. The seam had increased from 8

inches at the outcrop to 20 inches at the end of the pit, a distance of say 6 feet, or about 2 inches to the foot. The coal appeared to be of good quality, and may yet prove thick enough to be valuable. The outcrop of this seam was noticed in many places along the face of the ridge, but is not known to have been cut at any other place, unless the opening made by *Mr. G. F. Gaither* in section 4 of this township, heretofore alluded to, is on this seam. That it is, seems very probable.

Its position and surroundings correspond better with this seam than any other. The difference in the thickness of the seam at the two openings, one 20 the other 36 inches, may be altogether due to the depth to which the respective openings have been cut into the seam.

Neither of them had penetrated far enough into the hill to make a fair and satisfactory test—the capping rock and roofing slates were not reached—only enough exposure was made to render it reasonably certain that this is an important and valuable seam of coal.

The *Bynum Seam* No. 27 of the *General Section.*

This coal seam was opened by Elijah Bynum, near the middle of S. 17, T. 12, R. 3 E., on the southeast side of the quartzitic ridge. This seam is about 180 feet above the *Phillips* or *Lower Baine* Seam, and just above the fourth conglomerate rock, some of the pebbles of this rock are found in connection with the seam. It has as yet been only partially opened, and accurate measurements could not be made, but it shows approximately the following

SECTION :

Shale and soft sand rock.................20 feet
Roof, blue slate......................... 4 feet
COAL, very good......................... 4 feet 4 inches
Under clay.............................. 3 feet
Soft sand rock and conglomerate..........30 feet

When examined, the opening or pit was partly filled up, and only the upper part of the coal seam was visible. It was described by the man who opened it, as having a 2-inch slate parting near the bottom; approximately the seam carries about the following

SECTION:

Coal, solid, good...............................36 inches
Slate parting................................... 2 inches
Coal, solid, good... .:.........................14 inches

Thickness of coal at outcrop, 50 inches.

There is good reason to anticipate an increase in the thickness of this seam when cut in beneath a solid roof. It would require a further cutting of ten feet at this place to get under a solid cap rock, Until that is done, a fair exposure of the size of the seam, or of the quality of the coal it carries, cannot be obtained.

This coal has been freely tested in blacksmith forges, and is commended as a good shop coal. The seam is evidently one of the best in this coal field; probably carrying a thicker body of solid coal than any other yet discovered.

But unfortunately this seam is of but very limited extent. So far as yet known it only underlies portions of sections 17, 18 and 19 of T. 12, R. 3 East. Possibly a little of this seam may exist in section 9 of this township, though its outcrop was not seen there, and this opinion is based solely on the fact that a portion of that section is apparently high enough to contain it.

The space which this seam is certainly known to underlie is about 700 acres. But small as this area is, should the seam only hold the thickness it shows at the outcrop, it contains over 5,000,000 tons of coal.

The strata in the ridge containing this seam are nearly horizontal. This ridge stands just where the northwest dips of Blount Mountain meet the southeast dips from Straight Mountain, both gradually declining into the horizontal

Coal mining could be cheaply done here. Self draining tunnels could be driven into this seam on either side of the ridge, at any desired point, thus avoiding the usual heavy expense of pumping water and hoisting coal.

The outcrop of this seam was traced from the Bynum opening in section 17 toward the southwest. It shows very plainly in places around the irregularities of the ridge to the southeast ¼ of section 18, where is found the southwestern outcropping of the *fourth conglomerate*. About 10 feet above this rock and little farther to the northeast is the outcrop of the Bynum Coal Seam, its farthest southwestern margin. From this point toward the northeast the ridge rises about 50 feet higher than the outcrop of this seam for a half mile, thence for the like distance is a further rise 25 to 50 feet. On the northwest side of this ridge the outcrop of this seam is generally very plainly discernable, just above the upper edge of the decomposed conglomerate. It is not marked by any bold outcropping cap rock, as most coal seams are marked. The cap rock of this is evidently soft friable coarse sandrock, which only occasionally shows on the surface at all, and never prominently.

Towards the northeast corner of section 17 the ridge becomes very narrow, and it is cut by a gap in the southwestern ¼ of section 9, called the Hayse Gap, in which the strata is abraded down below the fourth conglomerate. No evidence of this seam was found any farther to the northeast, though in portions of this section (No. 9) the ridge again rises higher and the large pebbles of the fourth conglomerate are often seen in abundance, and not always on the highest ground. Some parts of the northeast ¼ of this section and of the northwest of section 10, and of the southwest of section 3 of T. 12, R. 3 E., are seemingly high enough to contain remaining detached portions of this seam. When searched for, the horizon of the underlying conglomerate pebbles will be the best guide the prospector can follow in seeking its location.

No. 28 of the *General Section*, the top seam of this coal field, has no distinctive name.   Its position is about 30 feet above the *Bynum* Seam. or 35 feet above the top of the *fourth conglomerate*.   It has only been dug into in one place, where it was two feet thick.   This may not be average thickness for the seam,   Its outcrop was seen in several places, and its position satisfactorily obtained.   It does not appear to be well roofed in, and the coal is probably soft, neither does it occupy quite as much space as the Bynum Seam No. 27.

The measures above the Bynum Seam are mainly composed of irony shale and soft reddish clays, with very little rock appearing on the surface.   Yet there may be in places a sufficient roof of rock, or hard slate, to give this seam the necessary protection against the disintegrating agencies of nature, and make it of some commercial value.

Above Coal Seam No. 28 there are rarely more than 20 or 30 feet of strata.   The ridge which makes the top member of the measures in this field is smooth-topped, with surface gently undulating.   It rises gently near the divide between the headwaters of the Blackburn and Locust Forks, and close to *Straight Mountain*, forming on its southwestern end apparently a part of that elevation, but gradually separating from it towards the northeast.   Between this ridge and Straight Mountain rise the extreme headwaters of the *Calvert Fork* of the Little Warrior.   This stream flows northeast between this ridge and Straight Mountain to about the middle of S. 8, T. 12, R. 3 E., where it has cut through the latter, and pursues its southwest by west course down Murphree's Valley.

This ridge begins to rise higher towards the northeast, near the Range line between Range 2 and 3 East, and its rocks begin to show a distinctly quartzitic structure.   It probably attains its greatest elevation in S. 17, T. 12, R. 3 E., but there is no material diminution of its height to the Hayse Gap heretofore mentioned, in section 9 of this township.   It is again nearly cut through in section 3, same township, by the head stream of Whipporwill Creek, which rises near its southeastern side. Approaching the Locust Fork, its altitude is sensibly diminished by the declination of the strata in that direction; and that stream divides it near its northeastern end.

This ridge which is so prominent and important on account of the valuable coal seams it contains, is comparatively narrow, only about a mile in diameter at the base, and often less than half a mile broad at the top. It stands close to Straight Mountain for about half its length, there being a space of only from 100 to 300 yards between them, but they gradually separate and become farther apart toward its northeastern end, where they are nearly a mile apart. In other words, the northeastern end of this ridge is about one mile southeast of the vertical southeastern edge of Bristow's Cove; but there is a small vertical uplift *between them*, and within one-fourth of a mile of the base of this ridge. This vertical uplift is probably, as heretofore noticed, a branch, or fork, of the Straight Mountain uplift, which deflecting eastwards, is plainly seen extending as far as the northeastern end of this ridge, but beyond that is not perceptible on the surface.

The foregoing embraces all the observed facts and details of coal seams in this field, which may be regarded as reliable, or reasonably certain. There are other seams of coal the identity of which have not been conclusively settled, and matters of fact and observation which may be stated only hypothetically. All these are reserved for the next section.

Section V.

UNSETTLED QUESTIONS.

While much time and labor and study have been given to this field with a view of identifying every coal seam, and

clearly outlining its structure, yet owing to certain irregularities of level and of faulting, this has not in all cases been
clearly accomplished.

The greatest difficulty in identifying seams was found
along the southeastern base of Straight Mountain, where a
fault of varying and uncertain depth exists. This fault
partly exposes, but also dissevers the mountain from, the
northwestern edge of the productive coal measures. Several
important coal seams are found in close connection with this
line of fault which as yet can be only doubtfully referred to
their proper position in the field.

Among these doubtful seams is the *Waide seam* already
described under the head of *"Explorations"* (page 23). It
was doubtfully referred to No. 8 of the *General Section* under
the head of *Details*. That reference was made solely on the
similarity of the surroundings, the cap-rock and intervening
slates, the coal seam being under deep water, and unseen.
Could the coal seam have been examined at this, or other
places in the river where it was presumed to come near the
surface, and to be cut by the water in deep places, fuller
evidence of identity would probably have been obtained.
Usually the surroundings of a coal seam, the overlying
slates and cap-rock, afford better evidence of identity than
the size and structure of the coal seam itself. The latter
may vary from place to place in quality, thickness and partings, all produced from local inflows of sediment during the
period of its formation; the former resulting from a general
inflow of sedimentary covering, more widespread, uniform
and persistent, necessarily give characteristics of greater
uniformity. From these surroundings alone, and from the
further fact that no other seam of coal possessed similar surroundings in that part of the field where its outcrop must
necessarily come to the surface, this one was without much
doubt referred to the *Waide seam* No. 8. This would, if
accepted as sufficient evidence, fix its southeastern outcrop
in nearly a southeast course from the Waide Gap where it

was discovered, but does not serve to fix its position except inferentially in other parts of the field. In fact it is not yet certainly known that this seam has been seen at any other point than the Waide Gap. Yet it is very probable that at its proper horizon it is co-extensive with this coal field. A coal seam of about 3 feet thickness exists at the *Allgood Gap*, in S. 4, T. 13, R. 2 E., in connection with a large body of slates, which may possibly be this seam. But it is among the vertical measures, with no cap-rock or other surroundings to afford it identification. Should this be the *Waide seam* it is evident that its normal place is far beneath the surface, and can only be reached by deep mining. In the other gaps through Straight Mountain, between this one and the Locust Fork, its position is still deeper, and it is not brought to the surface by the vertical uplift. Its southeastern out-crop in the middle of the field is necessarily in the high lands which make the divide between the two branches of the Warrior, where all out-crops are so thoroughly covered up that no coal has yet been discovered.

The Cowden Seam—not classed and probably not included in the *General Section*—presents another case of unsettled identity. This seam, of which a description and sections are given under the heading of *"Explorations,"* was one of the first coal seams opened in this field. It was opened the first year of Prof. M. Tuomey's service as *State Geologist of Alabama*, under the direction of *George Powel* one of his assistants. The opening on it was made by John P. Cowden, who had discovered it, and is hence called *"the Cowden Seam."*

The assistant reported the seam as being 4 feet 10 inches thick, and Prof. Tuomey paid Mr. Cowden a premium of $10.00 for its discovery.

The first opening on this seam may have been made where it was abnormally thick; subsequent openings near the same place show a seam of 4 feet 2 inches thick, including a parting of 4 inches of soft slate, dividing the coal into two
5

benches; about two-thirds of it above the parting, and one-third below.

Other openings made presumably on the same seam, in gaps through Straight Mountain, yet farther to the southwest, show a still thicker parting with slightly less coal.

The character of this coal where first opened is peculiar. It is massive, rather dull in color, cokes imperfectly, and yields a rather copious red ash. Farther to the southwest the quality of the coal is somewhat improved, but yet it retains all of its distinctive features.

The unsettled questions about this seam are, what is its true position in this field, and what relation does it bear to the other seams? These questions being unsettled, it is not included *by name* in the GENERAL SECTION of this field, though it may be included *in fact.*

That *section* was constructed by *measuring across all the known out-crops of seams,* on the longest slope of the field, from southeast to northwest. If it is *not included* in that *section,* then its southeastern outcrop has not yet been discovered, and it must be in one of those wide spaces of measures which have hitherto seemed barren of coal. If it *is included,* then the character and appearance of the seam, and the quality of its coal, are materially different at its southeastern and northwestern out-crops. One or the other of these propositions is apparently true.

Of the first proposition, it may be said that it seems improbable that in the many streams descending from the top of Blount Mountain toward the river, a coal seam of the size of this one should not have been cut and exposed, if it existed in that space. And yet coal seams undiscovered may exist, as has been heretofore suggested, in that wide space that exists between Nos. 3 and 4, and also between Nos. 4 and 5 of the General Section.

That the Cowden Seam could not crop out between seams 3 and 4 may be considered demonstrated by its greaten distance than No. 4, above the second conglomerate. Seam No. 4 is about 300 feet above that conglomerate, while this seam is over 600 feet above that rock, as shown by the ex-

posed strata, beside the unknown amount of fault or upward sliding of the vertical rocks, between it and the conglomerate, as exposed in Straight Mountain. If no fault or
upward thrust of the strata existed here the measurements
would make the Cowden approximately coincide with the
Sand Creek No. 1, No. 5 of the *General Section*, and the coals
of these seams bear a close resemblance to each other. But
the seams do not correspond in thickness, nor in the general character of their surroundings, so far as these could
be seen on Sand Creek.

We must moreover consider the probable amount of upward thrust of the vertical rocks of Straight Mountain,
which certainly cannot be less than the whole height of the
mountain—from 300 to 400 feet—or at the lowest estimate,
than the difference between the height of the top of the
mountain and the general level of the adjacent coal measures. This would give an apparent depression, by so much,
to the measures, as compared to the Straight Mountain
rocks, hence they appear in this section *below* their normal
place in the field.   Taking this fact in connection with the
measurements made, it becomes evident that the southeastern outcrop of this seam must be looked for higher up in
measures than the upper Sand Creek Seam, and that its
probable position southeast of the Cowden opening will be
at, or near, the Blackburn Fork of the Little Warrior.

This portion of the field contains many coal seams, and of
these Nos. 7 to 14 inclusive of the *General Section* have been
cut by that little river, yet none of these carry similar coal,
or have the appearance of the Cowden Seam. And there
seems to be none there yet undiscovered.

Apparently the solution of the difficulty must be found in
the other proposition, that some one of the seams already
known has changed in character and quality of its coal, between its southeastern and northwestern outcrops. Which
one is it? The answer can only be given hypothetically.
The seam that from its position and general make-up would
appear the most probable one, is the *Big Holt Seam* No. 7
of the *General Section.*

The Holt Seam carries about 4 feet of coal and nearly 2 feet of black band, but with an aggregate of over 5 feet of foreign matter. This foreign matter, clay, slate and rock, were probably *local inflows* of sediment during its formation, and hence not co-extensive with the seam. It is from this cause that partings in coal seams are produced, and they are for the same reason subject to great and frequent variations in thickness, often terminating altogether.

This *Holt* or *Big Seam* is probably the equivalent or counterpart of the "MAMMOTH SEAM" of the *Cahaba Coal Field*, which shows unusual variations in its splits and partings. Both occupy about the same position in the coal measures, both are about the same size, over 12 feet from roof to under clay, both have rocky parting near the middle. The Mammoth Seam splits into two seams toward the south, may not this also split toward the northwest? Or may not the partings diminish in that direction and all but one disappear as it is in the Cowden Seam? These querries can only be answered satisfactorily by future extensive explorations and practical mining operations.

A coal mine is now being opened on this seam in S. 4, T. 14, R. 1 E., near the bank of the Blackburn Fork, about one-fourth of a mile above the railroad bridge, which, in its progress, may throw additional light on the relations of this very interesting seam of coal. At this opening its position is about 15 feet above the bed of the river, with a dip of 3 deg. to the southeast; this dip in one-fourth of a mile will put it below the bed of the river which here flows northwest with a fall of 10 to 15 feet to the mile. The tunnelling of this seam for any considerable distance will afford data for calculating its range, dip and southeastern outcrop.

The seam known in the Cahaba Coal Field as the *Brock Seam*, lying just above the first or *Lower Conglomerate*, has not been opened or cut anywhere in this coal field, yet its existence is plainly indicated at many places by fossil coal plant impressions in the rocks at this horizon. And it was seen, and had been dug into, yet farther to the east, near the top of Chandler's Mountain, lying above and close to the Lower Conglomerate or Millstone Grit rock. Its thickness here was less than one foot, and it is probably a thin seam all over this field. From the fact that it was not cut or opened, it was not embraced in the General Section and is only referred to now to show the persistency of the seam, and as further evidence of the close relation in structure which obtains in different parts of the Alabama Coal Fields.

The quantity of available coal in this field is another of the unsettled questions. It has been clearly shown in the *Section on Details* and by the *General Section* of the field, that there are *eleven* (11), possibly twelve (12), coal seams in this field that are over *three* (3) *feet thick*. Ample thickness for advantageous mining. The uncertainty as to whether there are 11 or 12 of these arises from the uncertainty in the identification of the southeastern outcrops of the Waide and Cowden seams. Now it would have been an easy matter to estimate approximately the area of each of these coal seams, and to calculate and aggregate their solid contents. Yet all this would have served no practical purpose, and would have been wholly hypothetical.

It would have been based on the supposition that these seams had been cut and measured where they were of average thickness, and that they all maintained that thickness all over their respective areas. This would be highly improbable. No reliance could be placed even on the approximate correctness of calculations based on such uncertain data. They might greatly exceed or fall far below the actual amount. It is sufficiently shown in this report that there is in this field ample coal for *all mining purposes* and *for all coming time.*

Much of this coal was discovered and exposed by the Alabama Geological Survey, especially in the years 1891-92. It was the general expression of the citizens of that section of the country "that more coal had been developed in one season by the Geological Survey than had been done by all

preceding prospectors." For the first time the coal seams
were traced out and their relations shown.

Now, with the limited means and time of the survey devoted to this work, it is not probable that a full development of any one of the coal seams was obtained. Work had
to be stopped on every opening before the full thickness or
best quality of the coal had been reached. Only in one or
two openings was any penetration made beneath the cap
rock, and in every instance the coal seam was sensibly thickening when the work was stopped. Hence any estimate
based on the measurements of seams thus obtained would
necessarily be too low. Nothing but practical mining on
each one of its important coal seams will disclose the full
value and importance of this coal field.

It is shown that the productive measures at the thickest
part of this field have a volume of over 3,400 feet, and that
the sub-conglomerate measures have a thickness of at least
800 feet, making the whole thickness of this coal formation
over 4,200. It must, however, be understood that this immense thickness of coal measures is to be found only adjacent to the Straight Mountain.

This part of the coal field is an unsymmetrical synclinal
basin with its axis close to its northwestern edge, hence the
deepest part of the measures lie adjacent to the mountain
above named, and they thin out gradually towards the southeast, or towards the top of Blount Mountain overlooking the
great Coosa Valley.

Only the great industrial developments of the future will
fully expose this important coal field.

SUMMARY OF CHEMICAL ANALYSES.

By Dr. J. M. Pickel, University of Alabama.

COALS.

Number.	1	2	3	4	5
Moisture	1.49	1.27	1.46	1.17	1.18
Volatile combustible matter	32.38	36.49	32.11	34.90	32.35
Fixed carbon	61.46	56.19	64.97	59.52	64.18
Ash	4.67	6.05	1.46	4 41	2.29
	100	100	100	100	100
Coke	66.13	62.24	66.43	63.93	66.47
Sulphur	1.79	3.87	3.87	2.20	0.92

No. 1. Peacock Seam.
No. 2. Haynes Seam.
No. 3. Armstrong Seam.
No. 4. Woodward Seam.
No. 5. Baine Seam.

[THE END.]

·INDEX.

ERRATA.

On page 21, last line, for "*Bucks' Pocket*" read "*The Dungeon.*"
Page 23, eighth line from bottom, for Southeast read Southwest.